Why Safety Cultures Degenerate

From Chernobyl to Fukushima, have we come full circle, where formalisation has replaced ambiguity and a decadent style of management, to the point where it is becoming counter-productive? Safety culture is a contested concept and a complex phenomenon, which has been much debated in recent years. But it is not so much about what can be measured and pin-pointed as about what is difficult to fully articulate. In some high-risk activities, like the operating of nuclear power plants, transparency, traceability and standardisation have become synonymous with issues of quality. Meanwhile, the experience-based knowledge that forms the basis of manuals and instructions is liable to decline. In effect, future generations might interpret these instructions literally. In the long-term, arguably, it is the cultural changes and its adverse impacts on co-operation, skill and ability of judgement that will pose the greater risks to the safety of nuclear plants and other high-risk facilities. Johan Berglund examines the background leading up to the Fukushima Daiichi accident in 2011 and highlights the function of practical proficiency in the quality and safety of high-risk activities. The accumulation of skill represents a more indirect and long-term approach to quality, oriented not towards short-term gains but (towards) delayed gratification. Risk management and quality professionals and academics will be interested in the links between skill, quality and safety-critical work as well as those interested in a unique insight into Japanese culture and working life as well as fresh perspectives on safety culture.

Dr Johan Berglund of Linnaeus University holds a PhD in Industrial Economics and Management from KTH Royal School of Technology, Stockholm. His thesis 'The New Taylorism' addressed the safety culture of the Nuclear Power Industry and the issues around safety and quality in high-risk activities, as well as cultural changes and imperatives of working life in general. As Visiting Scholar at Meiji University, Tokyo, he had the occasion to develop these perspectives further, exploring the background of the Fukushima Daiichi nuclear accident of 2011. Prior to completing his dissertation Johan published a number of papers on the Nuclear Power Industry in Sweden, a collaboration between KTH and the Swedish Nuclear Safety and Training Centre (KSU).

Why Safety Cultures Degenerate

And How To Revive Them

Johan Berglund

Routledge
Taylor & Francis Group

LONDON AND NEW YORK

First published 2016
by Routledge

2 Park Square, Milton Park, Abingdon, Oxfordshire OX14 4RN
52 Vanderbilt Avenue, New York, NY 10017

Routledge is an imprint of the Taylor & Francis Group, an informa business

First issued in paperback 2020

British Library Cataloguing in Publication Data
A catalogue record for this book is available from the British Library

Library of Congress Cataloguing in Publication Data
A catalog record for this book has been requested

ISBN: 978-1-4724-7606-7 (hbk)
ISBN: 978-0-367-60601-5 (pbk)
ISBN: 9781134765829 (web PDF)
ISBN: 9781134765898 (ePub)
ISBN: 9781134765966 (mobi/kindle)

Typeset in Times New Roman
by Out of House Publishing

Contents

Preface

Discussing the background leading up to the accidental events at Fukushima Daiichi in 2011, this book highlights the implication and function of practical proficiency in the safety and quality of high-risk activities. Addressing the significance of experience-based knowledge in a reliable safety culture, this implicates discussions on what way training and further education can support and nurture the informal learning of skilled practitioners, along with the kind of *reflexivity* required to energise risk awareness and critical thinking. In a technological culture like that of the Nuclear Power Industry, the pursuit of formalisation is strong. Yet the uses of technology, and the application and in-depth understanding of rules and instructions, rely on experience. In the long term, arguably, it is the cultural changes and their impacts on co-operation, skill and ability of judgement that will pose the greater risks to the safety of nuclear power plants and other high-risk facilities. With regard to safety culture and related aspects of Quality, the accumulation of rules and instructions is not enough. It can even be counter-productive. The learning that goes into skill must be nourished rather than impoverished. At worst, operatives and plant personnel are forced to reduce their judgements, alongside the reflective and critical thinking indispensable to develop and maintain a capability to intervene so as to manage the unexpected; in the long run this is the quality that make safety cultures robust, dynamic and *antifragile*.

Acknowledgements

The embryo of this book was spawned when I was asked by Professor Richard Ennals to publish a working paper at Kingston Business School, on the subject of skill and formalisation and the quality work of safety-critical organisations. Richard has also commented on different versions of the manuscript, and has been a great support. What has made this project possible, though, is the research funding of the Swedish Civil Contingencies Agency (MSB), whom I gratefully acknowledge for the support of my post-doctoral term. I will also like to thank my students and colleagues at the Linnaeus University, where I have been based during the last couple of years. Especially Bo Göranzon, who has developed an epistemological framework and methodologies for practice-oriented research, who also commented on the writings for this book. In the context of Linnaeus University, Caroline Dunne has commented on the final draft of the manuscript. I would also like to kindly thank Thomas Lennerfors, as well as Professor Aki Nakanishi at Meiji University, who accepted me as visiting scholar in autumn of 2014 and helped me to organise and carry out research activities in Japan. Finally, I will also like to give a special thanks to my publisher.

1 Introduction: Skill and Formalisation

This book concerns areas of risk and civil contingencies, along with issues of quality, skill and learning. My objective is to establish a link between the safety and quality of high-risk activities on the one hand and experience-based knowledge on the other, a link that, I will argue, is crucial in the build-up and maintenance of a reliable safety culture. My key subject of investigation is the Fukushima Daiichi accident of 11 March 2011, the most severe nuclear accident since Chernobyl in 1986. In a wider context, I have looked into why safety cultures degenerate and how they can be revived. As the learning that goes into skill are associated with local and social contexts, it has been necessary to consider various background factors, broadly-based issues, which are likely to have an impact on other issues of importance, while not necessarily being the more obvious focal point of analysis; for instance underlying cultural factors or long-term cultural changes of work. In other words, wider implications of the Fukushima accident that go beyond the technical aspects of the events that have been described in detail in numerous reports, considering issues of significance across organisational borders, and to society.

Thus, nuclear operations remains one of the most notorious examples of high-risk industry in terms of its potential for catastrophe and with the politics and psychology that surrounds it; it is in our consciousness. There is considerable secrecy attached to this sector of industry and for that reason we tend to regard it as an entity of its own, despite the various similarities it has with other sectors of industry. Accidents and incidents relentlessly put in question what is often referred to as the safety culture of these high-technology systems and facilities. Subsequent to the misfortunes at the nuclear plant Forsmark 1 on the east coast of Sweden in 2006 – where the defence-in-depth reactor safety system did not operate sufficiently, due to a short circuit in a 400 kV switchyard outside the plant which affected other levels of the facility in unexpected ways – this accident was described as "the culmination of a long-term decline in safety culture".[1] Ultimately, operatives at Forsmark were able to untangle the situation without any harmful effects. This was not the case, however, with the accident at Fukushima Daiichi, where huge amounts of radioactive material were emitted into the environment.[2]

The 15-metre tsunami that hit the east coast of Japan on March 2011 disabled the cooling of three Fukushima Daiichi reactors, as all three cores melted in a matter of a few days, causing a severe nuclear accident. After a couple of weeks, the three reactors were by and large stable with water addition. While there have been no recorded deaths from radiation so far, for decades to come this area of Japan will be effectively uninhabitable, whereas up to this point more than 150 000 people have been evacuated. At this time it is impossible to say when this area will be habitable. Evidence also reveals that many non-human organisms, from plants, insects and birds to monkeys, have been significantly impacted by the radioactive releases in the form of genetic damage and high mortality. In addition, within an area of 20 km up to approximately 300 km from the destroyed power plants, there has been various other effects, on the ocean as well as the entire ecosystem, related to the Fukushima disaster.[3]

The strong requirements for safety within high-risk activities have triggered a greater level of formalisation, to make modes of operation standardised into "tasks", creating a larger basis for organisational knowledge. Despite that, experience-based knowledge remains a substantial constituent of any dependable safety culture, as rules and instructions must be understood and applied in real-time situations of practice that are not simply dependent on any predefined goals or tasks to fulfil. From encountering a great variety of situations, skilled practitioners seem to develop what can be characterised as the skill of anticipation. This, in turn, requires acquisition of "readily available, tacit knowledge", a capability to act in situations that are undetermined.[4]

Safety culture is a contested concept and a complex phenomenon. Organisations, much like societies, are created by humans, in which knowledge is developed in social and situational contexts. Besides, in high-risk activities there is an inclination to equate learning with positive changes in observable behaviour, as against fixed standards. Within the Nuclear Power Industry there is strong reliance on formal schemes of production, and on standard procedures, and like many industries it is now facing the challenges of a major generational shift in both workforce and technology. In addressing these challenges, manuals and instructions have been modelled for every possible situation; literally thousands of pages of regimentation and systems of coordination that are becoming larger and larger, and more and more predominant. How is this a problem?

As has been established in the study field of professional skill, human professionals learn through examples and analogical thinking between situations that are similar but *non-identical*.[5] From this perspective, learning is rather the outcome of reflection upon real-time events and experiences. In the wake of nuclear accidents like Three Mile Island and Chernobyl, there have been strong pressures on utility companies to develop more comprehensive work procedures. Many of these accidents have also pointed the finger at (the notion of) the fallible human. Then again, trying to delimit the human factor through the introduction of new technology in the work place, or persistent demands for formalisation, will raise other concerns; the proficiency of skilled practitioners run a long-term risk of being undermined, not given enough opportunity for learning; to exert their ability of judgement, for reflection and critical thinking.[6] For that reason, professionals in various fields of practice can be constrained by larger factors.

The fact that well-experienced personnel frequently are able to untangle a variety of unforeseen situations, which sometimes occur in high-risk facilities, makes this a matter of great urgency. Erosion of skill, in connection with generation shifts, extensive formalisation, as well as other long-term cultural changes of work, must be taken into account in conjunction with the phenomena of risk and safety culture.[7] A major concern in Sweden has been that of an ageing workforce, especially with regard to plant personnel. Other aggravating circumstances include the decision taken in the early 1980s to phase out nuclear power by the year 2010. This has been postponed, bringing into concern the issue of ageing facilities. Another main concern internationally has been the increase of output and revenue in addition to persistent demands of cutting costs on operation and maintenance. In other words, to re-design production systems to maximise efficiency.[8] Accordingly, safety might not be the only priority to the management of high-risk organisations.

This book is based on the following empirics: (1) *The Official Report of the Fukushima Nuclear Accident Independent Investigation Commission*, NAIIC, released in the summer of 2012, and related reports regarding the background factors of these events; (2) as visiting scholar at the Meiji University in Tokyo in the autumn of 2014, I got the opportunity to interview people from, or with great insight into, the Japanese nuclear industry. I also had the opportunity to visit the site of Fukushima Daiichi; (3) the broad empirical material on the Nuclear Power Industry in Sweden that has been generated since 2008, in collaboration between the KTH Royal Institute of Technology and the Swedish Nuclear Safety and Training Centre (KSU); (4) the launch of the new Master's Programme in Skill and Technology at Linnaeus University, the basis of which being humanities, epistemology and the study of experience-based knowledge; an educational initiative that can confront some of the issues elaborated in this book.[9]

The methodology that has been used is primarily hermeneutical, where dialogue and exchanges between researcher and informants takes place over longer periods of time, during which I have had my interpretations validated and rectified. The method that has been used to illuminate key issues of skill, quality, safety and education within this sector of industry is the Dialogue Seminar Method,[10] through which reflection is qualified through dialogue in smaller groups. Each participant prepares for a seminar by the reading of a shared impulse text as well as the writing of a short reflection based on their own experience. As the participants are able to articulate more of which they "did not know that they knew", the objective of such a seminar series is what could be outlined as a sort of discovery of *praxis*. Experiences are not precisely the same from person to person, but they can be illuminated and *developed* by various people interactively; discovering something together with a key role for dialogue and analogical thinking. In other words, we have applied all these carefully thought-out performance tools and systems of risk assessment, not to mention a proper organisational design: But what is it that we actually *do*?

Typical of the last 15 years or so is the revival of Taylorism in disguise, so to speak. This "new" Taylorism, as it is suitable to label it, has spread all through society and reveals itself in attempts of complete regimentation of experience-based knowledge by means of instructions, audits, models and table sheets; in insistent demands on scientific-like knowledge and formal education; in the measurement of performance and proficiency; in re-engineering of work; in a narrowing of vision due to persistent demands on increased "efficiency" and profitability. This is what the New Taylorism is essentially about. As an ideology it is less coherent, less openly declared and, compared to historical Taylorism, arguably less conscious; it is "at once more general and more restricted".[11] Also, its measures of control are more subtle, yet nonetheless effective, with immense impacts on modern-day work place organisations.

The exploration of human skill and what methods can be utilised to support experience-based knowledge are essential to this study. From this viewpoint, the analogical and critical thinking of operatives and plant personnel, gathered from hands-on experience as well as training, becomes an essential constituent of dynamic safety cultures.[12] Revolving around the Fukushima Daiichi nuclear accident, the objective is to arrive at some measures of practical application, which can help to improve and broaden quality work and the uses of training and further education in high-risk activities; to prevent technological risk, general deteriorations of safety and quality. When we talk of degeneration we usually refer to the state or process of decline; in terms of biology, degeneration usually means some sort of evolutionary decline, or loss of function, for instance in an organism or a species. In physical degeneration certain functions in cells or tissues are reduced, impairments which *can* be reversible. The same may well be true with regard to safety cultures and the quality of work place organisations in general, in which case degeneration is likely to transpire in the form of a worsening of moral qualities, or other qualities and faculties that characterise a certain group of people or a culture. When something degenerates it gets worse by some means. Likewise, degeneration is generally undesirable, a sort of antithesis to development and incremental change; from the Latin word *degenerare*, it originally means "to be inferior to one's ancestors".[13]

History of course knows the processes of development and degeneration and the various forms they can take. Yet the notion of degeneration is not always neutral. It can also be biased. Arguably, the best way of clarifying such corrosive processes is by means of examples from different sectors of industry, which is what I will be trying to do. The key issue, however, and arguably the biggest challenge, would be to recognise these long-term processes of decline or gradual deterioration before they become virtually irreversible. This, I hope, is among the things that this book will help to illuminate.

Notes

1 Cf. *Background: The Forsmark Incident 25th July 2006*, published by the Analysis Group at KSU (The Swedish Nuclear Training and Safety Centre), and Larsson, L. and von Bonsdorff, M. (2007): *Ledarskap för säkerhet (Leadership for Safety)*, an independent report discussing the developments leading up to the incident/accident at Forsmark in 2006.

2 On the day of the accident three of the facility's six reactors were operating at full power whereas the others were shut down for maintenance and refuelling. For a more detailed event summary of the Fukushima Daiichi accident, see INPO, "Lessons Learned from the Nuclear Accident at the Fukushima Daiichi Nuclear Power Station" (2012): pp. 6–7.

3 "Biological Effects of Fukushima Radiation on Plants, Insects, and Animals", www.phys.org/news, 14 August 2014, accessed 10 August 2015.

4 Sennett, R. (2008): *The Craftsman*, pp. 172–178.

5 See Göranzon, B. and Florin, M. Eds. (1992): *Skill and Education: Reflection and Experience*. The focus of this research area has been on case studies, examples and analogies that highlight something, or some aspects, of the tacit knowledge of skilled professionals within different areas of expertise.

6 See Göranzon, B. (2009) [1990]: *The Practical Intellect – Computers and Skills*. This has been established independently in case studies in both Sweden and Japan in the early 1980s.

7 My PhD Thesis, *Skill and Formalisation* (KTH Royal Institute of Technology, 2011), explores these issues from an epistemological position. It discusses the uses of training and the role of experience-based knowledge in the safety cultures of nuclear power plants.

8 See Berglund, J. (2013): *Den nya taylorismen – om säkerhetskulturen inom kärnkraftsindustrin. (The New Taylorism – On Safety Culture)*. PhD Thesis, revised edition.

9 Within this educational initiative, experience-based knowledge can be explored in an academic context. In this interdisciplinary study field, the division between research and education is less accentuated than within many other areas of research. The research field of skill and technology, accordingly, is not about the design of new technology, or of work place organisations for that matter, but about *application* of various technologies. In this context, seminars and student papers will provide fresh perspective on issues of skill and tacit knowledge within various areas of work.

10 See Göranzon, B., Hammarén M. and Ennals R., Eds. (2006): *Dialogue, Skill and Tacit Knowledge*, pp. 57–65. In addition, I have made study visits at the recurrent training of plant personnel at Swedish power stations Forsmark 3 and Ringhals 3/4.

11 Cf. Doray, B. (1988): *From Taylorism to Fordism*, p. 155.

12 Snow, C. P. (1998): *The Two Cultures*, p. 62, delineates the concept of culture as intellectual development, development of the mind, or "development of those qualities and faculties that characterise our humanity", a nation, or a certain group of people. As for safety cultures, this can translate into organisations, facilities or sectors of industry.

13 See www.vocabulary.com/dictionary/degeneration, accessed 8 August 2015.

2   The Separation of Knowing and Doing

2.1   The New Taylorism

In the early twentieth century, American engineer Frederick W. Taylor developed the theory of scientific management, anchored in utility maximisation and standardisation of work. From a Taylorist point of view, the maximum efficiency of work could be accomplished only through enforced standardisation, by means of detailed instruction developed by specialists or management experts; ensured by scientific, or *scientific-like*, procedure. In that sense, a higher level of understanding and efficiency is equated to a higher level of codification and regimentation. Taylor believed in transferring control, and thinking, from the factory floor upwards.[1] For the feasibility of scientific management it was a dilemma of sort that the mass of experience, the collective knowledge of the workers, were not in the hands of the managerial expertise, who then would be more qualified to assess and evaluate which mode of operation would be the most efficient, or adequate, for each task:

> Under scientific management the "initiative" of the workmen (that is, their hard work, their good-will, and their ingenuity) is obtained with absolute uniformity and to a greater extent than is possible under the old system [...] The managers assume, for instance, the burden of gathering together all of the traditional knowledge which in the past have been possessed by the workmen and then of classifying, tabulating, and reducing this knowledge to rules, laws and formulae which are immensely helpful to the workmen in doing their daily work.[2]

Taylor acknowledged the power of evolution and some of his writings were coloured with Darwinian-like discourse and metaphors. However, he saw what he thought was a better way to attain quality, namely through management interventions by means of scientific methods and scientific-like approaches to practical knowledge. In the course of my enquiries regarding the Swedish nuclear industry it was also apparent that enforced standardisation had significantly increased. People in various parts of the industry also expressed concerns with this tendency.

From established errors, incidents and "near misses" gathered from commercial power plants all over the world, formal procedures and "best practices" are implemented into local practices. Commonly known as Operating Experience, collection and revision of experiences advocated by intergovernmental authorities like the International Atomic Energy Agency (IAEA), and collaborative organisations such as the World Association of Nuclear Operators (WANO), has a learning agenda; a kind of formalised experience that is authorised to nuclear utilities worldwide through benchmarking. By means of observation and questionnaires, these accredited bodies also carry out various forms of auditing processes, collecting explicit data to evaluate; to certify organisational behaviour appearances and standards of operation, as well as the fulfilment of rules and regulations.[3]

Monitoring of performance through operational safety audits has the objective of assuring that the necessary constituents for safety have been implemented and institutionalised, in other words, that they "comply with pre-specified and auditable criteria" as required by legislation, regulation or international best practice.[4] The question is: Does all this comparing, benchmarking, evaluating and auditing also contribute to the promotion of *learning*?

Typically, emulations of best practices are imposed on operating organisations through management, so as to improve operational safety and effectiveness; but also to receive higher rankings in the WANO Peer Review, or the IAEA Operational Safety Review Team (OSART), typically aiming at the upper quartile level of WANO safety leading indicators; creating a strong incentive for these management standards to spread. Nevertheless, these developments also possess certain risks. If the safety of industrial power plants is equated to the compliance with rules and standard procedures, facilities such as these, as political scientist James C. Scott has argued, would in fact be *less* effective:

> (The) formal order encoded in social-engineering designs inevitably leaves out elements that are essential to their actual functioning. If the factory were forced to operate only within the confines of the roles and functions specified in the simplified design, it would quickly grind to a halt. [...] The more schematic, thin, and simplified the formal order, the less resilient and the more vulnerable it is to disturbances outside its narrow parameters.[5]

This also delineates some of the shortcomings of Taylorism and enforced standardisation: In "Taylorised" work places the abstract is separated from the concrete. The system of dividing up work, by means of identification and differentiation of tasks, provides a basis for the continuous observation of labour and of the workforce. Through the quantification of output, managers are more easily able to compare various groups or individual workers with others, to supervise as well as to control them. Once the "norm" has been established, performance is evaluated by the extent to which the output or behaviours matches the norm. The work process itself thus becomes increasingly fragmented and mechanised.[6] In the words of Bernard Doray, who has analysed the long-term implications of Taylorism and Fordism in the French manufacturing industry:

> In that sense, we can explain the vitality of Taylorism by looking at what it conceals as well as what it helps to reveal. Certain of Taylor's errors are highly significant in that, predictably enough, the complexity of reality undermined the utilitarian results he hoped to obtain. [...] It is important to do so, because the influence of Taylorism on a whole range of contemporary scientific practices takes the form of a reduction of the field of knowledge.[7]

A century after its introduction, due to the attractive package it represents to people who are keen on taking a direct route to "efficiency", the schemes of Taylorism are still on the agenda. Yet its theoretical tools and approaches to human work have not led to any greater understanding of human skill and labour. In many areas of work, extensive formalisation has rather created a growing gap between what people know and what they are actually empowered to do or say. In that sense, the issue of safety culture has a more general implication.

American sociologist Richard Sennett has argued that the kind of rocky nineteenth-century capitalism that inspired the theories of Karl Marx, on labour as a source of alienation, has in recent decades undergone something of a revival. The strong emphasis on *flexibility* of modern-day working life has generated new types of uncertainty. Within many companies managers are more or less obliged to streamline their organisations on a regular basis. This has become the general way of showing investors that the company is to be reckoned with.[8] Typically, large consultancy firms are brought in to push and legitimise the downsizing and re-engineering of enterprises and organisations, creating "a decisive break with the past"; reducing staff along with the number of approved methods and modes of operation. Still, many such efforts fail and have often also proved to be counter-productive, as "business plans are discarded and revised; expected benefits turn out to be ephemeral; the organisation loses direction".[9] The stability that once was has been replaced with a growing uncertainty.

In the mid-1990s, corporations such as Wal-Mart, as well as Japanese manufacturers like Toyota, Nissan and Honda became benchmarks of a new economy. Their enterprise systems and productivity numbers have been promoted by economists as well as global management consultancy firms who specialise in re-engineering, where consultants by and large are trained in streamlining businesses of all sorts.[10] Today, re-engineering of work has found its way into all types of professional activities, even medical care:

> The goal is to standardize and speed up medical care so that insurance companies can benefit from the efficiencies of mass production: faster

treatment of patients at reduced cost, with increased profits earned on increased market share.[11]

This can be exemplified by the wide-ranging remodelling of the British National Health Service (NHS) over the past decade or so, which has been subjected to New Public Management-style reforms. With the objective of modernising the nation's health and medical care system, the people in charge of the NHS implemented the so-called *Ford Model*, an example of management by numbers, which revolutionised American car industry in the early twentieth century; taking the centralised control to an extreme, performance and output are measured in terms of targets that are entirely quantitative.[12]

Technically, this means that doctors are reviewed by the number of tumours or cirrhosis they have treated, and the amount of time spent with each patient, rather than how many *patients* were successfully treated. In other European countries a number of reports in recent years have pointed to a widespread discontent among doctors and nursing staff, while their skills in dealing with patients are being "frustrated by the push for institutional standards".[13] In the end, the experience-based skills of doctors and nursing staff are effectively marginalised. Combined with the widespread use of contractor companies, this has also meant that very little experience is actually shared by the various stakeholders in between.

In other words, there is this commanding albeit chivalrous notion of an increasing body of knowledge being produced centrally, for organisations all over the country to learn from, to absorb and utilise. The current debate in Scandinavia points to a similar assumption, seeing that rationalisations of medical care have led to various quality gaps, and there have been similar discussions regarding the public educational system. Essentially, doctors and nursing staff have had to spend more time on non-core activities like administration and less and less time with patients; this is also relevant to other areas of the public sector such as schools, universities and the police. Re-engineering is often based on thin simplifications of complexity and the repetition of familiar themes. Guided by a single vision, these approaches to quality and efficiency prefer the direct over a continuous process of discovery and adaptation. However, in hindsight it is easier to distinguish just how many attempts of re-engineering, on a social or organisational level, have also failed in achieving its intended objectives. Their anticipated connections between intentions and outcomes were false; they were based on an oversimplified view of the world.[14]

The knowledge ideal of modern society is theoretical and intimately related to the concept of a *model*, claim Gustafsson and Mouwitz at the National Centre for Mathematics Education in Gothenburg, as abstractions delineated with the support of one or a small number of examples:

A model is perceived as *general*, i.e. it claims applicability to a variety of new situations, it should be able to approximate reality, and be applicable to the complexity of specific cases in the future, as well as able to explain or forecast concrete future events. A model is explicitly formulated [...] to be able to be communicated through education. Since the model should be able to explain the complexity it has been extracted from, some practical complications arise. In many cases, successive adaptation between the model and the individual case is required for its application to be possible. Sometimes a real situation creates such intensive "resistance" that the model must be revised or rejected. If the problem has to be resolved quickly, the model must be replaced by the hands-on knowledge and skills of the labour force, as when an unanticipated error suddenly occurs, for example, with a nuclear power facility.[15]

This is in many ways typical of the knowledge ideal of the Nuclear Power Industry and it is also a fitting description, as implied, of the knowledge that goes into manuals and instructions. Whereas models are conceived as all-embracing, practical proficiency has another texture; it is of a more analogical type. These analogies consist of a number of situations and examples connected with each other, as each new situation is compared and related to previous examples and experiences, gradually building up a proficiency, or skill, which has "a more or less *general applicability* without claiming the generality of a model".[16]

As for medical care, doctors and nursing staff often have to diagnose from insufficient evidence of a patient's condition, and in many cases narratives are fragmental or inconclusive. To attain an overarching picture with regard to various symptoms, as well as a patient's lifestyle and catamnesis, is bound to take time, experience and judgement. An experienced doctor is not far from a *detective*, who, working with loose ends, "when the clues are confusing" or scarce, ultimately has to decide on a proper treatment:

> [...] we most need a discerning doctor when a diagnosis is not obvious, when the clues are confusing, when initial tests are inconclusive. No simple technology can serve as a surrogate for the probing human mind.[17]

In Taylorised work places, this is the type of skill that is liable to decline. This is relevant also to the skill and ability among operatives of nuclear power plants, within which accidents are sometimes the cause of minor, independent faults whose consequences interact simultaneously between different levels of such high-technological systems. When facing the unexpected, plant personnel are sometimes forced to take quite creative measures to handle such a course of events, aggravated by insufficient or contradictive information.[18]

The supply of experiences and observations of skilled practitioners provide a presence of the unexpected in everyday practice (that is the skill of anticipation), making such expectancy a permanent state of mind. In other words, given the proper conditions for learning, the human factor is a potentially strong link in a reliable safety culture: the refined capability to manage uncertainty is vital to the skill that operatives within this type of industry accumulate over time. The day-to-day supervision of the process, and of the facilities, is more intricate than what is sometimes assumed. For that reason, the human ability of learning to see patterns and to detect minor changes, in for instance a certain process, becomes vital in the operational safety of high-risk facilities.[19]

Over recent decades, management control has taken new forms. Numerous standards for risk management are being produced across various sectors of industry, creating new demands for proof and evidence of action.[20] As in the nuclear industry, the highest quality is gradually replaced by the *right* quality. Adaptation to the more and more costly processes of certification and auditing, to a large extent an expanding and "self-preserving" structure, has thus become next to synonymous to issues of quality and legitimacy within professional activities such as schools, hospitals and universities. This, subsequently, will require some sort of *reflexivity*, self-evaluation or readiness to the fact that sometimes there might not be any true need for such arrangements:

> Reflexivity will require an institutional confidence to dismantle as well as construct audit arrangements. Regulatory sensitivity about what makes organizations like schools and hospitals effective is necessary, a sensitivity which involves decisions about how to leave individuals alone to get on with their work as much as how to monitor them. This in turn will require recognition of the manner in which practices are perpetuated isomorphically because they have become legitimate and not necessarily because they have been even moderately effective in achieving goals.[21]

When Taylorism is mentioned today it is usually in relation to the far-reaching measures of streamlining the work tasks of highly qualified personnel. This is often accompanied by a short-term push for organisational change and profit maximisation. In an era of deregulation and fierce competition, this New Taylorism has found its way into work place organisations of all sorts. Apparently, in many professional activities a "hidden" Taylorism has crept up on us, as with the re-engineering of medical care in the US and several European countries, in which the objective has been to speed up the average time doctors spend with patients, making each member of staff invest less time with each patient, but also to re-design work itself, directing the very contents of these consultations. Many times, managerial agendas and rationalisations like these have an opaque albeit strong administrative impact.[22] Through these cultural changes, hence, new variations of Taylorism are spawned; new varieties on the separation of thinking and doing.

In his early writings Marx had a desire to set the modern craftsman free in order to realise the development of the individual and the value of work; to reawaken the spirit of craftsmanship and the craftsman as a "maker of civilisation". To the Soviet leaders, in contrast, the principles of Taylorism and scientific management made for a more attractive package in ensuring the superiority of the Communist system. These principles would also saturate its educational system. Nevertheless, the productivity of the Soviet Union remained considerably low in the decades leading up to its collapse. On the part of the authorities there was distinct disbelief in, or even fear of, local self-management of factories and husbandries. Hence, in these top-down utopias there was a sort of emptiness in the "collective, moral recipe for craftsmanship".[23] Japan, on the other hand, has had its own command economy with a great deal of government guideline and ideology. They have influences of Taylorism but the cultural imperative to work well is strong, which has manifested itself in both productivity and practical creativity.

So, when we talk of a *Japanisation* of working life in recent decades, are we in that case witnessing a revival of Taylorism in disguise? As implied, after the Second World War, the concept of scientific management had a strong impact on Japanese management styles and theory. Many enterprises emulated the basic ideas of Taylorism, and its successor Fordism, while hierarchy governed the Japanese work place. In the 1960s and onwards this was supplemented by the introduction of Quality Circles and other small-group improvement-type activities, fostering a higher level of cooperation, empowerment and greater involvement among workers in various sectors of the organisation. Rather than continuing along the lines of the Taylor system of management by specialists, where some do the thinking while others simply do what they are told to do, many corporations were looking to strengthen their existing organisations by engaging more members of staff in problem solving, problem finding and in quality control, giving each employee a greater opportunity for learning. This "learning style" was also a means of improving the potential for innovation, encouraging the input and self-initiative of workers. These developments also had a long-lasting influence on Japanese management literature and vice versa.[24]

Japanese corporations at the time also had another vital influence, that of the business analyst W. Edward Deming, who advocated that managers should "get their hands dirty"; that the empowerment and collective craftsmanship of organisations were to be fashioned by "sharp mutual exchanges":

> Within the Japanese factories it was possible to speak the truth to power, in that an adept manager could easily penetrate the codes of courtesy and deference in speech to get across the message that something was wrong or not good enough. In Soviet collectivism, by contrast, the ethical as well as technical centre was too far removed from life on the ground.[25]

Today, a similar thing is happening in many organisations, in Japan and other countries as well. Contrary to the prospects of Taylorism, in professional activities there is rarely a "one-to-one match" between means and ends. Most people have an innate capacity to become skilful within a certain area of expertise. The problem today, according to Sennett, is that western society in general does not sufficiently explain to the individual person that he or she has a long-term potential to learn far beyond the context of formal education. Young people today are rather encouraged to secure a portfolio of various *skills*, whereas skill in a sense of craftsmanship is based on "slow learning" and habits, which "establish a rhythm between problem solving and problem finding"; providing a strong basis for dealing with reality.[26]

While many western economies have basically stopped growing, alongside the persistent stagnation of Japan, economists around the world are desperately looking for new paragons.[27] As in the case of Japan, for many young people there has been a shift in opportunities over the last 20 years or so. Part-time and temporary workers, with fewer career opportunities and little training, have increased from one-fifth to one-third of the labour force. Being closed off for so long, Japanese society was comparatively easy to keep homogenous, and in that way to keep everyone working together; a lot of the economic successes after the Second World War may well be attributed to the hard work of that generation. Working within traditional company structures people had job security, as well as a definite idea of their place in society. As a result, many Japanese industries became strong and famous. Once the economy collapsed in the 1990s, for many people that level of security disappeared. Still, younger people are told that if they are successful in the educational system they will have the same level of security, but that is not the reality any longer. Consequently, people are starting to question society a lot more than they used to.[28] Also, in recent years it has become evident that the number of highly skilled workers is decreasing. Within Japanese industry, reliant on what can be characterised as industrialised craftsmanship, this is becoming a major problem.

2.2   Managing the Unexpected

On 11 March 2011, the tsunami caused by the Great East Japan Earthquake of the east coast of northern Japan, one of the largest earthquakes in recorded history with magnitude 9.0, flooded and destroyed the emergency diesel generators, the seawater cooling pumps, the electric wiring system and the DC power supply for Units 1, 2 and 4 of the Fukushima Daiichi nuclear facility, run by the Tokyo Electric Power Company (TEPCO). Apart from an external supply to Unit 6 from an air-cooled emergency diesel generator, this resulted in an overall loss of power. Most Units lost power, but the flooding did not damage only the power supply. The 15-metre tsunami also destroyed buildings, equipment installations and other machinery, while seawater from the tsunami flooded the entire building area. All in all, the loss of electricity made it difficult to cool down the reactors, also resulting in the sudden loss of monitoring equipment such as scales, meters and other control functions in the control room. Lighting and communications were strongly affected. Unlike the neighbouring facility of Fukushima Daini, the off-site power supply of Fukushima Daiichi was cut off, due to damage to the transmission towers from the earthquake, and the supply from backup batteries was not enough to cool all reactors. Ultimately, decisions and responses to the accident had to be made "on the spot by operational staff at the site, absent valid tools and manuals".[29]

Despite these extreme events, Chairman Kiyoshi Kurokawa of the Fukushima Nuclear Accident Independent Investigation Commission (NAIIC) claims that the Fukushima accident is to be regarded as a *man-made*, rather than a natural, disaster. Despite its excellence in engineering, the Commission points at some ingrained conventions of Japanese society, counter-productive to the ability of managing the unexpected: "our reflexive obedience; our reluctance to question authority; our devotion to 'sticking with the programme'; our 'groupism'; and our insularity".[30] In addition, there was inadequate evaluation as to the risks of rare events with small probabilities, such as tsunamis.

Interestingly so, this in-depth critique of Japanese culture is touched upon in the English preface only, and not in the original report. The Commission typically blames the incapacity in Japanese mentality for critical thinking, and to admit failure and ambiguity. In addition, Japan's nuclear industry did not in any effective way absorb the lessons learned from the accidents at Three Mile Island and Chernobyl.[31] In other words, there has been strong meta-criticism within Japanese society in recent times, but arguably also a great deal of denial.

According to The Institute of Nuclear Power Operations (INPO) Independent Review, with participation from WANO, a lot of the actions taken to respond to this accidental course of events suggest that on-shift personnel had the required level of proficiency. In this improbable and stressful situation, many staff members showed great resilience and personal initiative in their numerous efforts to restore critical safety functions, preventing the accident from becoming even worse. Then again, there were also "knowledge weaknesses" that could be traced to "practices that were not developed using the systematic approach to training process" (SAT).[32] Yet such approaches to safety and quality are directed towards the management of the expected, not necessarily the unexpected.

As the accident went well beyond previous experience, operators and emergency response personnel on the whole responded well, despite the fact there was insufficient communication between the Head Office and management on site. Besides, strong reliance on computer-based training for instance as regards "accident management", as well as infrequent recurrent training of plant personnel, is likely to have created vulnerabilities in knowledge retention as well as in-depth understanding. This review, one must take into account, is characteristic of the kind of discourse typical for supervisory authorities and collaborative organisations. Accordingly, TEPCO, along with other Japanese utilities, are criticised for having missed out on opportunities to improve the decisive ability to withstand flooding, and emergency responses in general, by not making broader use of international best practice and operating experience. For instance, to consider what other factors might have resulted in the same consequences in relation to other accidents. On some accounts, INPO are also appealing for increased automation. In the vein of the NAIIC report, it is also recognised that within these organisations there seem to have been an absence of critical thinking:

> In retrospect, TEPCO would have benefited from additional questioning and challenging of the assumptions that a large tsunami capable of flooding the plant could not occur. Additionally, questioning and challenging of assumptions may have helped maintain core cooling during the Fukushima event when communications were difficult and reliable information on plant parameters was unavailable.[33]

In other words, critical thinking is vital to the maintenance of a dynamic safety culture.[34] Furthermore, there must be channels for turning this critical thinking into a "questioning attitude", the challenging of assumptions and *status quo*. To make matters more complicated, several accidents also seem to result from a series of decisions that reflect flaws in the "shared assumptions, values, and beliefs" of operating organisations, including those at Three Mile Island and Chernobyl. According to INPO, in the past ten years or so TEPCO had taken a number of actions to strengthen several aspects of the organisation's safety culture, such as their quality management system, for instance by adopting a "corrective action program"; it had also developed a set of safety culture principles, based on WANO best practice, with a broader communication of all reported difficulties that had safety culture implications. Other practices put in place to promote and monitor nuclear safety culture included a safety culture performance indicator with the purpose of continuously tracking the trends of safety culture. Even so, they seemingly failed in finding the right balance.[35]

Constance Perin, researcher at MIT, has distinguished three logics of control, or cultural dimensions of safety that regiment matters on American nuclear plants: The *calculating logic* of deduction, assessing various levels of risk related to high-technological systems; the *real-time logic* of plant personnel, extracted from local practice, countervailing any deficiencies in construction, or technology in a wider sense; the ability to sense, interpret and monitor the array of signals, alarms and messages given out by these systems. Thirdly, the *policy logic* directs the decision-making concerns and processes of management, in trying to find a balance between production targets and the safety directives given out by the regulatory authorities. According to Perin, there is an "inherited caste system of credibility" within this sector of industry; a downward spiral from "mind" to "hand", in which the calculating logic and to some extent the policy logic set the standards and procedures of practice. The credibility problem derives from the fact that the "experience-distant", quantitative logics of engineers and risk management has high status and a well-developed terminology, while the "experience-near", local knowledge of plant personnel is to a large extent *tacit*:

> To calculate risk levels for reactor designs, engineers move through a series of deductions to shape high-level control concepts into risk estimates, then into the algorithms, thresholds, and standards that become the design basis, the technical specifications, and the procedures and rules for maintaining control. Once reactors are operating, real-time logics reduce and handle risks that design calculations have not anticipated.[36]

From the perspective of the NAIIC, all these logics of control were in some way inadequate. Engineers are usually schooled in calculating and eliminating risks, but similar to other high-technologies, process plants do not simply behave as predicted. In other words, the unexpected cannot be ruled out. System upsets and disturbances outside the margins of safe operation occur from time to time, and operating experience is not always transferable between various organisations. If facing the undetermined, or when normal conditions no longer prevail, control room operators have little choice but to draw on memories, observations, experiences and judgements in order to cope with discontinuity, instabilities, anomalies or other dubious misadventures. Likewise, disturbances outside the narrow parameters of the calculative approach have to be interpreted and accommodated by the "hierarchically inferior" understanding of practitioners operating at the sharp end of the organisation:

> Even though model-makers themselves may be aware of the limits of their simplifications, models may find themselves doing more heavy lifting than intended when the world goes its own way. What model language calls "ill-structured" problems or "system upsets" are more likely to be real-world situations made more intractable by overstructured expectations that a technology will and should perform as modelled.[37]

When operational decisions are made in situations like these they are not likely to be based on rational analysis, but on the basis of the information at hand in a given situation; a skill-based choice between perceived alternatives for action. The "heavy lifting" of models and their thin simplifications, or even retrospective distortions, of reality is more reminiscent of a production of comfort and credibility, rather than practical solutions to the ill-structured problems of real-world situations.[38] In that way, the separation of "knowing" and doing is not always as rational as it may appear.

In some contexts of course the opposite also applies; activities in which knowledge based on personal experience is most influential. Rasmussen and Svedung (2007), in making a case for increased levels of legibility and transparency with regard to high-risk organisations, highlight the court reports of the train collision at Clapham Junction in 1988 and the Moura mine accident in Australia in 1994. In case of the latter, the mine company was under high pressure to deliver coal to a new power plant, pressures which influenced the direction of management, ultimately leading to a major explosion. As the court report discloses, information acquired by word of mouth was considered more important than written guidelines. Moreover, there was a style of management with a tendency to dismiss any contrary evidence, and managers did not pay much attention to reports from the levels below. They simply took for granted that all the necessary information had been conveyed informally to minors and deputies. As for this local work context, there was no consistent safety culture due to the weaknesses of communication as well as the indecisive commitments of management.[39]

In other words, procedures and channels of communication were either insufficient or too informal, which underlines the importance of finding the right balance. There has to be some sort of overarching structure of operation and cooperation, while leaving a certain degree of freedom for skill- and experience-based judgement and decision making. And, what is perhaps more interesting is whether operatives and other decision-makers at the sharp end were for some reason inhibited in their ability to intervene so as to manage these accidental courses of events. Replacing a decadent style of management with excessive formalisation is arguably not the way forward, in which case the learning from everyday practice is likely to decline; communication and cooperation based on dialogue impoverished.

According to external plant evaluations of INPO, the Fukushima Daiichi facility was in immaculate condition, up to the accident in 2011, although there had been some problems at other power stations run by TEPCO. The risks with regard to future earthquakes and tsunamis were known to chief executives, but on the whole considered highly improbable. The facilities were well-organised, with well-maintained equipment. They had great performance records and excellent housekeeping, but were still not prepared for the worst. The focus of training was predominantly on the expected, and there was little readiness for worst-case scenarios. Also, there was a culture of compliance typical of the Japanese, where employees were not encouraged to constructively disagree with superiors.[40] Though there was also a culture of continuous improvement activities, like Quality Circles, it arguably was not enough to untie the Japanese aversion towards disagreement, the challenge of assumptions and authority.

When we speak about risks we usually mean negative events or incidents, which are in some way conceivable, that we do not know for sure will happen, but *can* happen. Risk can be estimated in terms of probabilities, but also in terms of the potential for disaster.[41] What would be considered normal or acceptable during the rocky, early period of nuclear power would be deemed unacceptable in our time due to new and higher demands of both safety and revenue, and with the more direct approaches to quality pursued today.

All technologies are conjoint with risks and as risk analyst Nassim Nicholas Taleb has pointed out, the "improbable", or "unexpected", is never well-defined; concepts like these are not steady, but dependent on the

individual or group making the judgements. What we are able to anticipate and prevent rely on experience, and what we have experienced in life.[42]
Moreover, the notion of an event and the classification of specific types of
events is to some extent elusive. That is, "the more accurate the definition of
an event, the less probability that it will ever be repeated",[43] and the rarer the
event, the less we know about the frequency of its occurrence.

> Man-made complex systems tend to develop cascades and runaway
> chains of reactions that decrease, even eliminate predictability and cause
> outsized events. So the modern world may be increasing in technological
> knowledge, but, paradoxically, it is making things a lot more unpredict-
> able. [...] the odds of rare events are simply not computable. We know
> a lot less about hundred-year floods than five-year floods – model error
> swells when it comes to small probabilities.[44]

Specifically, the methods of analysis used for internal event risk are more
developed and have "smaller associated uncertainties than those used to assess
the risk of low-frequency natural-phenomenon hazards".[45] For that reason,
Taleb challenges the notions of risk and resilience: the resilient operator or
organisation responds, and recovers, but remains the same. Instead we ought
to discuss, and evaluate, whether for instance a certain system is fragile, or
antifragile; whether it becomes vulnerable, or on the contrary gains strength,
from disorder and volatility.[46] Due to the feeble prospects of calculating the
risks of rare events and small probabilities, and predicting their occurrence,
we want safety cultures to be as dynamic or robust as possible, not so much
resilient but "antifragile"; to benefit from errors and stressors, and to improve
when exposed to discontinuity and uncertainty relative to a given situation.
I shall come back to this later.

More than anything, it is the claim of science that has made Taylorism
popular in various parts of the world. When, in recent decades, the separation of thinking and doing has re-emerged in new forms, it is usually in the
pursuit of safety, control, "efficiency" and quality. Yet it is the practitioners
that make the key decisions at times of uncertainty, and in novel situations
and circumstances, and although they have a huge amount of manuals and
instructions to match to a given situation, as in the case of the Fukushima
Daiichi accident, they may also have to act on their own sense of judgement
and experience. In that sense, quality work also implies looking at what people actually do, rather than imposing on them models of how economists or
engineers think people should behave.[47]

Notes

 1 Taylor, F. W. (2007) [1911]: *The Principles of Scientific Management*, pp. 30–31.
 2 Ibid., p. 34.
 3 Cf. Rasmussen, J. and Svedung, I. (2007): *Proactive Risk Management in a Dynamic
    Society*, pp. 73f. Bodies like WANO and IAEA usually have slightly different

priorities and questionnaires when evaluating. Questions put forward could be of the following type: What are the responsibilities of each person within the organisation? What performance tools are used to monitor human action?

4 Woods, D., Dekker, S., Cook, R., Johannesen, L. and Sarter, N. (2010): *Behind Human Error*, p. 79.

5 Scott, J. C. (1998): *Seeing Like a State*, p. 351.

6 Doray, B. (1988): pp. 40–41.

7 Ibid., p. 85.

8 Sennett, R. (2006): *The Culture of the New Capitalism*, p. 78.

9 Sennett, R. (1999): *The Corrosion of Character*, p. 49.

10 Head, S., "They're Micromanaging Your Every Move", *New York Review of Books*, 16 August 2007, p. 42. See also, Head, S., "Inside the Leviathan", *New York Review of Books*, 16 December 2004, and http://www.accenture.com/Global/Services/By_Industry/Retail/Client_Successes/RetailProgram.htm, accessed 10 May 2015.

11 Head, S. (2007): p. 42.

12 Sennett, R. (2008): p. 47.

13 Ibid., p. 46.

14 Kay, J. (2011): *Obliquity – Why Our Goals are Best Achieved Independently*, pp. 7–9.

15 Gustafsson, L. and Mouwitz, L. (2008): *Validation of Adults' Proficiency – Fairness in Focus*, pp. 18–19.

16 Ibid., p. 19. As I have encountered, in most cases technical manuals and instructions are protected by secrecy.

17 Groopman, J. (2009): "Diagnosis: What Doctors Are Missing", *New York Review of Books*, 5–18 November, p. 26. The basis of the article is the book by professor of medicine Jonathan A. Edlow, *The Deadly Dinner Party and Other Medical Detective Stories*, Yale University Press, 2009.

18 Perrow, C. (1999): *Normal Accidents: Living with High-Risk Technologies*, p. 27.

19 Perby, M-L. (1995): *Konsten att bemästra en process* (The Art of Mastering a Process), pp. 195–196.

20 Power, M. (2007): *Organized Uncertainty – Designing a World of Risk Management*, pp. 1–3ff.

21 Power, M. (1997): *The Audit Society – Rituals of Verification*, p. 145. Power also makes a distinction between *auditing*, or an auditing style of monitoring, oriented towards control, normative proof and evidence of action, and *inspection*, or an inspective style of management, focusing on evaluation and *substance*. However, as in the case of extensive formalisation and long-term cultural changes, divisions like these are sometimes obscured and difficult to discriminate. See below.

22 Cf. Groopman, J. (2009): pp. 27–31.

23 Ibid., p. 29.

24 Stewart, J. R., "The Work Ethic, Luddities and Taylorism in Japanese Management Literature". *Industrial Management*, Nov 1, 1992, pp. 23–26.

25 Sennett, R. (2008): p. 31.

26 Ibid., p. 9.

27 Cf. Ferguson, N. (2014): *The Great Degeneration*, pp. 4f.

28 "Social Change in Japan: When the Myths are Blown Away", *Economist*, 19 August 2010.

29 Kurokawa, K. et al. (2012): *The National Diet of Japan – The Official Report of the Fukushima Nuclear Accident Independent Investigation Commission*, NAIIC (Executive Summary), p. 14.

30 Ibid., p. 9.

31 Ibid., p. 20.

32 INPO, "Lessons Learned from the Nuclear Accident at the Fukushima Daiichi Nuclear Power Station" (2012): p. 30.

33 Ibid., p. 34.
34 According to Woods, D., Dekker, S., Cook, R., Johannesen, L. and Sarter, N. (2010): pp. 78f, in so-called High Reliability theory, reliability is not the same as safety. Rather it is the dynamics or interactions of different parts of socio-technical systems that make things safe or unsafe. In other words, reliability "addresses the question of whether a component lives up to its pre-specified performance criteria", while organisational reliability is often associated with "a reduction in variability", and an increase in processes and behaviours that produces the same predictable outcomes. Following the line of argument presented in this study, robust or dynamic might thus be more appropriate in epitomising safety cultures that can manage uncertainty, or respond to shifting conditions.
35 INPO, "Lessons Learned from the Nuclear Accident at the Fukushima Daiichi Nuclear Power Station" (2012): pp. 33–34.
36 Perin, C. (2007): *Shouldering Risks – The Culture of Control in the Nuclear Power Industry*, p. 198.
37 Ibid., p. 203.
38 Ibid., p. 265.
39 Rasmussen, J. and Svedung, I. (2007): pp. 63–64.
40 *Presentation on the Fukushima Daiichi Accident in March 2011* by Wade Green, former Department Manager at INPO, at a Vattenfall workshop at Arlanda, Sweden, 6 March 2013.
41 Cf. Spiegelhalter, D., "Quantifying Uncertainty", in Skinns, L., Scott, M. and Cox, T. Eds. (2011): *Risk*, p. 17.
42 Taleb, N. (2010): *The Black Swan: The Impact of the Highly Improbable*, pp. 43–46; 51–55.
43 Rasmussen, J. and Svedung, I. (2007): p. 30.
44 Taleb, N. (2013): *Antifragile: Things That Gain from Disorder*, p. 7.
45 Corradini, M. and Klein, D. (2012): *Fukushima Daiichi – ANS Committee Report*, p. 23.
46 For a presentation of the resilient actions and interventions of High-Reliability Organizations (HRO's), see Weick, K. and Sutcliffe, K. (2007): *Managing the Unexpected: Resilient Performance in an Age of Uncertainty*, pp. 3f.
47 Kay, J. (2011): p. xii.

3 The Shadows of Progress

3.1 Functional Autism

The report of the NAIIC can also be traced back to the Japanese implementation and evaluation of the so-called *Information Society* in the 1970s. When in 1972 the term Information Society was first used in a Japanese futurological study, and the prospects of the computerisation of working life flourished, there was widespread belief that Artificial Intelligence (AI) would have replaced all manual knowledge by the year 2000. Yet in 1985 evaluators had begun to notice characteristic adversities in these developments, and the term *functional autism* was invented to pinpoint this typical decline of ability: "the phenomenon of people who work for a long time in a computerised environment, with its characteristic view of reality falling into categories of black/white, right/wrong, experiencing difficulties in confronting reality".[1]

In the early 1970s, a key intermediate target in the Japanese Plan for Information Society was the establishment of so-called *computer mind* among all citizens. By educating people in the use of computers in a variety of professional activities, the capacity of human professionals was expected to reach new levels of "highly intellectual creativity". In formalising knowledge into easily accessible computer-based information, a key target was the prevention of various "disastrous problems" that may arise in the near future. The national plan for the establishment of Information Society, it was assumed, would also bridge future gaps of demographic changes, for instance such as heavy shortages of intellectual personnel. The transformation of society, from industrialised to information-based, would include the creation of a *third sector*, financed by the government, but then operated by private enterprises, promoting public interests. In the implementation of this plan for social development, most efforts were to be concentrated to this third sector of society.[2]

Such comprehensive approaches in designing future life may sound naive, even though it has of course somewhat been realised today through the internet. In the goal towards year 2000 it was also assumed that in a future society people will be able to gain knowledge more quickly with a "computer mind", which was theoretically the prior key to success in the realisation of Information Society. Regarded as an intermediate target in the Computerization Committee report, the development of the computer mind, to be established by the mid-1980s, was not only assumed to improve the quality of society as a whole, but also more specific areas such as the progression of health and safety. Moreover, computer-oriented education was to be introduced at all levels, from primary school onwards. In comparison to the developments in the Nuclear Power Industry today, another objective was to measure educational effects more efficiently in realisation of a "standard educational system". It was thus considered important to speed up the process of learning through education, as well as to pinpoint its general effects. In modern society intellectual creativity would bear fruit in managing various sorts of problems more efficiently. Also, in the establishment of a computer mind, an intermediate step was understood to be the development of "information-oriented" behaviour, with extensive training in computer technology.[3]

Apart from the establishment of computer-based education, the other mainstay of social computerisation was the modernisation of the public health care system. To begin with, a first aid medical information system was to be developed. After that, it was intended to introduce computerisation in all general hospitals, which would include automation of hospital office work as well as increased efficiency in diagnosing and medical treatment. Other areas of society that were to be computerised would include supermarkets (food supply), pollutions prevention and the traffic system. Naturally, there were expected demerits of such radical mechanisation of work, but such disadvantages were considered relatively small compared to its expected merits to most sectors of society. There was some apprehension of decline, still, like a general loss of ethics and humanity.[4] This, too, was to be counteracted through a series of educational projects, creating a new "high-satisfaction society", in which the well-being and intellectual creativity of each individual would be highly improved.[5] In other words, the Japanese Plan for Information Society in the 1970s could also be seen as a sort of predecessor to the concept of Information Society or "Knowledge Society" in western society, and in our time.

At an early stage, there was even this idea of a new class of people, a kind of "techno-economists" who would operate this new society, using computerised information and high-technologies for the benefit of the general public. The biggest cost, however, for the promotion of Information Society and its intermediate targets, was supposed to be the implementation of computer-based education; an estimated total of 65 billion dollars, supplied by the Japanese Government between 1970 and 1985. Directed by government, the characteristics of Information Society were to be oriented towards the knowledge industry. While computerisation in welfare-related areas like medical care and education would be more and more advanced, the idea was to create a wide variety of expert systems on the basis of AI.[6]

As for the educational system, the idea was that things like traditional school years and traditional tests would be effectively abolished. Clearly, the analogy between the human mind and the computer was an important part of this educational vision, while major changes would include a shift from mass education to individual education; from passive education to self-studies, in which the computer and the information it holds becomes the "general knowledge".

In medical care, a man–machine diagnosis and treatment system would be developed and gradually expanded to cover all major cities in Japan by the early 1980s. Endorsed by a social re-education system, a man–machine conversational mode of interacting with the computer would presumably lead to a more rational and purposeful way of life and thinking, in which people would no longer act and behave simply in accordance with their feelings or intuitions. With the backup of the computer, the human ability to solve problems, it was so assumed, would increase significantly.[7]

This major Plan for Information Society was by and large the Japanese equivalent to the Apollo Project, the Japanese endeavour being social development instead of space exploration, while a lot of funding was to be invested in the development of new technology. This was meant to be a total refurbishment of society, even though there were some worries with regard to the creation of a "controlled society", with a strong governmental influence. For this reason some sort of sub-system, reflecting the opinion of the people, would have to be built into the total system of Information Society.[8]

The microelectronics revolution paved the way for high-speed, high-accuracy transferring of information. When in 1985 the Plan for Information Society was evaluated by the National Institute for Research Advancement (NIRA), according to their Electronics Research Group its prospects were in many parts accurate, that is that the establishment of computer-based work and education would transcend the limits of man's capabilities. The problem was that it had some serious adverse effects on human skill and proficiency.[9]

Surely, implementations of new technology have been crucial to the modernisation of society. Apart from the exaggeration of scientific-like knowledge and technology of our time, Japan has skilfully preserved a lot of its old cultural spirit, distinctively criticised in the preface to the NAIIC report; the lack of individual initiative within Japanese social systems in general. In its obsession to compete with the west, Japanese society has been both open (accepting the innovations of new technology from others) yet insular (somewhat reluctant to share its research results with the international community). Interestingly so, some of the more severe criticism of Japanese culture put forward by the NAIIC was accentuated by NIRA in the mid-1980s.[10]

The phenomena of "functional autism" relates to feelings of inability and alienation, of professionals working to rule, "jumping to hasty conclusions" rather than making considered judgements in situations of ambiguity; a kind of mental strain characteristic of people working for a long time in a computerised environment. There are other consequences still, like the risk of isolation while cooperative labour is weakened, the "formation of harmonious inter-personal relations and close cooperation" typical of Japanese factories and work place organisations.[11] There are other analogies too, to the current situation within the Nuclear Power Industry:

> If we ever fall into the trap of believing that technology can solve everything, we will end up discarding individual responsibility, a very high-touch value. Our illusions about the potential of technology account for this belief. We tend to think that technology will free us from all individual restraints and responsibilities.[12]

The notion of functional autism corresponds to the kind of hollowing out of skill and ability delineated in Swedish case studies of working life in the 1980s, most notably in the aftermath of computerised administration of social security offices nationwide. In this case, a decline in ability among more than 50 per cent of the administrative officials had occurred over time, regarding the fortitudes of decision making and personal judgement. One implication of the Japanese investigation was the need to counter-balance such long-term tendencies of decline within the educational system; another implication was the need to maintain the division within professional practices between *doing* something and writing a general rule for it.[13]

Tendencies such as these have also been detected within the Swedish nuclear industry, in which enforced standardisation have to a large extent superseded the "high-touch" value of skill and personal responsibility. The reason for this may well be the strong requirements throughout this sector of industry to demonstrate what means of control each organisation disposes of. Yet by increasing the number of rules, instructions and formal procedures, one does not necessarily enhance the overall *quality* of operations. At worst, relying on formal systems of codification becomes a way of escaping responsibility and critical thinking.[14]

In other words, the relationship between humans and technology is not trouble-free, and there are many forms of variability and complexity in the processes of power generation, health care or transportation. Consequently, determining more accurately what human functions and abilities should *not* be replaced by computers, comprehensive rule-systems, expert systems or other technologies, and which should be done so sensibly will be vital to success. Within the educational system, the development of technology and scientific-like principles of work must be promoted while preserving a balance between science and technology, and on the other hand the humanities and social sciences; to challenge rather than to consolidate various tendencies, or *shadows*, of progress. This was one of the closing remarks of the Japanese evaluation in 1985 and continues to be an issue of significance internationally – both in working life and in other areas of society.[15]

3.2   Technological Culture

The *Meiji Restoration* period of the late nineteenth century was a sort of Japanese equivalent to the Enlightenment, which accelerated industrialisation and modernisation, promoting western ideas and advances. Taking the developments of Europe and the US as model, the assumption was that social development could be "speeded up" by means of education and technology. During this period economic development and social change became a national goal, and official delegations were sent abroad on study tours to gain information of the political, economic and military systems of the west. In analogy, the Plan for Information Society in the 1970s was a way of exploiting western advances and taking it into the future; a second wave of modernisation and social development. But this, fundamentally, was confusing a humanistic concept of development with a technocratic conception of development, with a strong emphasis on the technical aspects of development, technical training and education.[16]

Professor Akira Omoto, a member of the Atomic Energy Commission of Japan, and with a background at both IAEA and TEPCO, who has investigated the Fukushima accident, also highlights cultural impacts as decisive, such as the lack of critical thinking, on both a national and organisational level, as well as within the nuclear community as such: Similar to the NAIIC, he points to an overall complacency, overconfidence and lack of professionalism due to a high level of outsourcing. Also, there has been strong *parochialism* within the nuclear community, and lack of dialogue; the priorities of risk management have been on cost-plus tariffs and the relation with local governors, not safety and continuous improvements.

In spite of the extreme events, according to this investigation the circumstances of this accident were not unique to Japan. Arguably, there are salient features of national culture that has contributed to the ultimate disaster: The Japanese "think and act as a group"; there is widespread collectivism, uncertainty avoidance and lack of independent challenge. This has also made its mark on the safety culture of specific power plants. Cultural attitudes are not easy to transform, but one of the significant issues of the Omoto study is also the backlog from modifications of the public educational system from the 1970s onwards and their long-term impacts on society.

In other words, certain aspects of a culture are liable to change, for instance through the educational system, making this a matter of relevance to other countries as well. In fact, there seem to have been some sort of overall loss of thinking capabilities within society as a whole; in Japan education has become a matter of transfer of knowledge rather than "learning how to think". Even if it was abandoned as a governmental strategy, this points in a direction towards the social development plan for Information Society, and the fact that it seem to have had some long-lasting impacts on Japanese society. In addition, Akira Omoto points to a culture in engineering with a tight emphasis on component quality and reliability, but with a lack of "big picture thinking", distracted by detail and formalism.[17]

This is an indication also that the implication put forward in the NIRA evaluation of the 1980s, to balance the shadows of technological change and the adverse consequences of a technological culture within the educational system, where not attended to. In the west there were equivalent movements, advocating the prospects of manual knowledge and labour being replaced by various sorts of (computer-based) expert systems. Anticipating a more and more sophisticated AI, this was assumed to pave the way for a democratisation of knowledge, for more efficient methods of learning and knowledge development. Hence, a vital element in working life was thought to be training in the new technology. Once the technology was mastered, the microelectronics revolution would allow people to "change from one area of work to another with ease".[18]

During this period, organisational theorists such as Ikujiro Nonaka were also planting the seeds of the Knowledge Management movement, the keystone of which has been the conversion of knowledge, from tacit into explicit, in conjunction with different methods of dissemination. When in 1950 the British mathematician Alan Turing introduced the field of AI in his paper "Computing Machinery and Intelligence", the question was: Will computers be able to "think" along the lines of humans? In other words, will digital computers be able to generate human-level intelligence? According to Turing, the prospects of this were favourable. Allied code breaking during the Second World War, in which Turing was involved, led to the development of the computer. Still, a computer would most certainly find difficulties in the formation of considered judgements on the bases of morals, human experience and sentiments. In the study of working life of today, rather, we try to figure out how we humans are fashioned by the interaction with computers and other technologies. Instead of looking solely at the developments of computer science, and the machine as a superior replica of ourselves, we must also consider the long-term effects of these man–machine interactions. Perhaps the real challenge of progress will be that humans are beginning to imitate the technology? That we will look at others and ourselves in a different manner, for instance when it comes to learning and proficiency, as the computer, not man, becomes the role model for human behaviour. If we can programme the computer to be like us, maybe we can train others and ourselves to be like computers, and to perform the same work? Today, when it comes to working life, we are communicating and interacting with each other through the computer and other technologies. For this reason we seem to be getting closer to people in other parts of the world, yet more distanced from the people around us.[19]

As established in the Japanese evaluation of Information Society in the 1980s, the attempt of establishing the computer mind had adverse implications on the forms of functional autism and erosion of skill. The analogy between knowledge and information turned out to be a fraudulent one;

something was lost in the man–machine conversational mode of interacting with the computer to solve problems more productively and reliably. In this case information became an impediment to knowledge, not so much a force of development as degeneration.

To Turing, computer programming was similar to creating a set of instructions, and the computer was similar to an individual who followed these instructions "by the book". In that sense, he was right: The computer is very good at solving problems that we have specified, less so in dealing with ambiguity. Planting the seed of AI, Turing seemed convinced also on the possibilities of building an imaginative computer equivalent to the human mind, which was the role model for AI. Our understanding of human performance, equally, would proceed in direction towards information processing. The philosopher Wittgenstein, in his account, dismissed such analogy. To him knowledge was only fully expressed in judgements. In that sense, clear-cut definitions of human concepts, or *rules*, are not sufficient, if at all possible. Whereas the computer repeats itself continuously, humans have a capacity for digressions, as their judgement evolves through experience.[20]

The mechanisation of work, the displacement of manual knowledge and the tendency to eliminate the human factor by means of automation have had similar far-reaching implications with regard to Taylorism and Fordism. In other words, the industrious rationalisation of human labour, the compulsive cycle of tasks to be carried out, has meant increased standardisation of the "human component":

> [...] to optimize the efficiency of the workers movements means that the innermost workings of the human body have to be viewed as though they are the workings of a machine.[21]

It seems that in a technological culture we humans are to some extent alienated from ourselves; we are more or less expected to learn and to function more like machines. The increased standardisation and mechanisation of work initiated through the implementation of scientific management saw the beginning of this and would inevitably lead to the dominant impression of "the body as machine", by which the estimation of the average speed at which workers carried out certain tasks and operations became a means of incorporating "the activities of men into that of machines".[22] Thus, in many workplaces the rhythm of skill and application was replaced by the notion of repetition of simple movements, entering the era of technological advances and imagination.

One way of transforming society is through the educational system. We know today that the expansion of public education does not automatically increase wealth, or economic growth. More than anything, this relationship is associative, not causal, and it may well be the other way around. But education can still increase equality in the population, "learning", as well as critical thinking.[23] The scientific process has the motive of understanding the natural world, as well as to control it. The humanities, on the other hand, has its primary interest in culture, self-reflection and in "the development of the mind". In the 1950s and 1960s scholars like C.P. Snow had begun to notice a growing division between the two "cultures" of the academic world and in western society as a whole. Engineers and scientists seemed to have less and less interest, for instance, in the writings of history, literature and philosophy, while people of a non-scientific background had little insight into mathematics as well as other sciences. In the aftermath of the Scientific Revolution this process now seemed irreversible, while the ideal of liberal education had begun to crumble away. Consequently, both sides were effectively "self-impoverished".[24]

The Dialogue Seminar Method was developed in such an endeavour, between 1996 and 1999, in collaboration between the KTH Royal Institute of Technology and the system development company SAAB Combitech, namely to bridge the gap between science and the humanities. What some managers at this company had realised was that many engineers were clever with detail and calculation but they lacked an overarching perception in dealing with problems of variability and complexity in situations of everyday practice. In shaping further education by creating a forum for so-called "experience development", these groups of engineers were able to tease out aspects of their proficiency previously hidden in practice. In the process they were encouraged to write down examples and difficulties from their own experience and in the course of dialogue jointly "develop" those experiences. Through analogies to literature and other fields of proficiency, fresh knowledge was illuminated; tacit knowledge for which there were no clear-cut language or models to describe it. Such experience development was thus a way of opening up for an expanded epistemology, to energise the experience-based, reflection-oriented learning of skilled professionals.[25]

This was also a way of departing from a system-based thinking, that the work process for new projects, such as the development of new products, could be standardised in order to make the end product fit the original specification; to find the one right way to serve all purposes. As for the revival of Taylorism, which has found its way into many areas of work, this has a general implication. To identify these developments, one may also look at how the conception of quality is produced and generated within safety-critical activities.

Given the results of the NAIIC, some of the deep-rooted conventions of Japanese culture were accentuated; a type of meta-criticism unheard of in many other countries. This report can also be traced back to the Japanese Plan for Information Society and its emphasis on formalisation of knowledge and work in the outcome of which the phenomenon of functional autism was pinpointed. From Chernobyl to Fukushima, have we come full circle, where formalisation has replaced ambiguity and a decadent style of management, to the point where it is now becoming counter-productive.

Notes

1 Göranzon, B. (2009) [1990]: pp. 139–140. Those futurological studies referred to are: Japan Computer Usage Development Institute: *The Plan for Information Society – a National Goal Toward Year 2000*, Computerization Committee, Final Report, May 1972, and National Institute for Research Advancement (NIRA): *Comprehensive study of microelectronics*, 1985.
2 *Plan for Information Society – a National Goal toward Year 2000*, Final Report, May 1972, p. 4.
3 Ibid., p. 34.
4 Ibid., p. 57.
5 Ibid., p. 88.
6 Ibid., pp. 68–69.
7 Ibid., pp. 109–128. Formalisation of knowledge, hence, is by and large equated to the development of skill and ability, by means of high-accuracy transfer of information, assimilated by computer mind as established through various educational projects. This is an example of education becoming more a matter of transfer of knowledge, rather than learning how to think. Still, we cannot know for certain to what extent and how the lingering effects of the establishment of Information Society, largely abandoned as a government project in the mid-1980s, has affected Japanese education, culture and society in the long run.
8 Ibid., p. 136.
9 National Institute for Research Advancement (1985): *Comprehensive Study of Microelectronics*, p. 118.
10 Ibid., pp. 11f.
11 Ibid., pp. 122f.
12 Ibid., p. 127.
13 Göranzon, B. (2009) [1990]: pp. 126f.
14 Cf. Berglund, J. (2013): pp. 71–102.
15 Göranzon, B. (2009) [1990]: p. 139f. Göranzon was one of the so-called *100 Wise Men around the World* consulted to give their views on the NIRA report in 1985, and on the establishment of Information Society.
16 Cf. Nordenstam, T. (1985): *Technocratic and Humanistic Conceptions of Development*, pp. 13–19.
17 Omoto, Akira, Presentation at IVA's Conference Centre, Stockholm, 29 November 2013.
18 Göranzon, B. (2009) [1990]: p. 136.
19 Sandblad, A. (2015): *Maskiner och människor* (Man and Machines), Master Thesis in Skill and Technology, Linnaeus University. Anders is System Developer at Combitech.
20 Ibid. Cf. Kay, J. (2011): pp. 46–47f.
21 Doray, B. (1988): p. 71.
22 Ibid., p. 65.
23 Taleb, N. (2013): pp. 203–206.
24 Snow, C. P. (1998): pp. 14f.
25 Cf. Backlund, G. and Sjunnesson, J., "Better Systems Engineering with Dialogue", in Göranzon, B., Hammarén M. and Ennals R., Eds. (2006): pp. 136f.

4   The Projection of Quality

4.1   Complexity and Rational Design

In high-risk industries standardisation is usually a sign of credibility, and extensive measures are often taken to monitor human action, illustrated by the fact that "the human factor" in many accident reports is regarded as the main source of error within socio-technological systems. Even so, time and again operators save technology when facing the unexpected.

The desire to make reality more legible and manipulable has partly to do with safety. It is also related to international benchmarking, as well as an increased focus on commercialism; a clear intention to link cultural values to certain market-related targets. This growing emphasis on commercialism can have various effects on different levels of organisations, and can also have consequences for the safety cultures of individual power plants, if taking focus from issues of relevance from a safety perspective among those in charge of the operation.[1]

Working in safety-critical utilities is not as "linear" as sometimes assumed, for instance when probing the intranet of high-risk organisations. The descriptions of models and standard procedures usually depict sequences of processes and actions during "normal" conditions. Yet reality is rarely free of disorder and variability:

> System operations are seldom trouble-free. There are many more oppor-
> tunities for failure than actual accidents. In the vast majority of cases,
> groups of practitioners are successful in making the system work product-
> ively and safely as they pursue goals and match procedures to situations.
> However, they do much more than routinely following rules. They also
> resolve conflicts, anticipate hazards, accommodate variation and change,
> cope with surprise, and work around obstacles, close gaps between plans
> and real situations [...][2]

Quite the reverse, incidents and accidents typically result from minor faults, or deficiencies, in one part of a system, interacting with other parts or units in unexpected ways, and caused by a number of irregular, coinciding events and conditions. When this occurs, operators are often faced with confusing beyond-design-basis events, in which they are obliged to take independent and sometimes quite creative measures to steer clear of danger; where there is "no single well-formulated diagnosis of the situation".[3] Charles Perrow has labelled such sequences of minor failures *system accidents*, or "normal accidents". Socio-technological systems, over time, have often proved to be more complex than initially assumed. In other words, time equals disorder and volatility. New devices are frequently added, but even if intended to pre-empt established faults or deviations, these amplifications often add to the complexity and *opacity* of systems, and can in themselves spawn new such accidents:

> (Technological fixes), including safety devices, sometimes create new accidents, and quite often merely allow those in charge to run the system faster, or in worse weather, or with bigger explosives. [...] We have produced designs so complicated that we cannot anticipate all the possible interactions of the inevitable failures; we add safety devices that are deceived or avoided or defeated by hidden paths in the systems. The systems have become more complicated because either they are dealing with more deadly substances, or we demand they function in ever more hostile environments or with ever greater speed and volume.[4]

This corresponds to the development in Sweden where all power plants have been upgraded for prolonged longevity as well as increased capacity. According to Perrow, this exaggeration of automation marks down something of a utopian mentality within many industries, that is to say that the facilities operating are consistently made safer with added technology. These types of upgrades are also implemented as a result of production pressures. Besides, the complexity of systems and the efficiency measures of management have often turned out to be an ominous combination.[5]

It is not always the operators' fault when things go wrong; technology can fail, something which plant personnel must be able to untangle. Yet with each added device, there is less and less room for manoeuvre and less opportunity for the learning from everyday practice, doubt and discovery. To inexperienced shift teams, the prospects of managing the improbable are often poor; to experienced operators system accidents are likely to appear more *linear* and thus manipulable. The knowledge that can be extracted from technical analysis of such events, still, is unlikely to be of very much relevance as to the detection and management of future system accidents. According to Perrow, even in the relative comfort of hindsight, so to speak, these are complex processes that can be to some extent "described, but not really understood", while they are usually perceived by means of trial and error, hands-on knowledge that has to a large extent reduced the number of major accidents. Even so, operators frequently get the blame for upsets and incidents of all sorts, including system accidents as the one at Three Mile Island in 1979.[6]

It seems implausible at the design and construction stages to forecast all future problems and contingencies that may occur when these technologies are actually *used*. Apart from the contingency of unexpected interactions of several abnormal coinciding events and conditions, there is more at risk. Except for the complexities of system accidents, there may also be cases where executives or top management ignore the repeated warning signals of their personnel, or other warnings, proceeding with unsafe practices.[7] Likewise, court reports from several large-scale accidents, such as the gas leak disaster at the industrial pesticide plant in Bhopal in 1984, or the nuclear catastrophe at Chernobyl in 1986, indicate that these were in fact the results of "a systematic migration of organisational behaviour toward accident under the influence of pressure toward cost-effectiveness".[8]

These large-scale accidents, in other words, were not primarily caused by the coincidence of minor independent failures or human errors. They were rather the cumulative effects of long-term cultural changes, in terms of a degeneration of safety culture. Regardless of what kind of migration is underlying these processes, disasters may well develop from a "tidy, well-ordered deterioration over a long period of time whose results would be described as disastrous if this deterioration took place at a specific time".[9]

Typically, the deregulation of electric power markets in the 1980s and 1990s have led to a concentration of power in the hands of a few major companies, similar to TEPCO who control both the production and distribution of about one-third of all the electricity in Japan. In addition, companies like TEPCO have a substantial influence on politics at all levels. To reduce costs, the Nuclear Power Industry also took its place on the "re-engineering bandwagon" beginning in the 1980s and continuing today to rationalise procedures and production systems to maximise revenue and efficiency.[10]

Competitiveness of markets, in general, creates a short-term incentive to increase profits, with measures such as outsourcing, downsizing and cost-savings on maintenance. In this case, it has also created competition between different energy sources. Anticipating the effects of deregulation on future profits, many power companies have adopted new strategies. While some companies have managed to improve operating efficiency, others were looking to cut operating costs by reducing maintenance costs and other expenses on a major basis, one example being the Millstone facility in Connecticut. Even though top management of these facilities were aware of the risks of taking deep cuts in current maintenance costs, their management consultants at McKinsey were called in to do the "justifying study".[11]

A few years later, in an internal memo, some of the consequences of these cost cuttings were exposed. In spite of increased profits, required inspections of facilities were not conducted and there was a failure to repair leaks and identify corrosion in cooling pipes that were not maintained. In addition, the US Nuclear Regulatory Commission (NRC) had noticed that whistle-blowers were often fired or mistreated for reporting safety violations, and in the mid-1990s the NRC had no choice but to, when further problems were discovered, shut down all three of the Millstone power plants, not allowing restarting without considerable improvements.[12]

This is a key example to illustrate how the priorities of executives and upper management can inflict a severe blow to the safety culture of high-risk organisations in terms of degeneration of quality. In recent times there have been similar problems at Swedish nuclear plants, whereas executives have been focusing on cutting maintenance costs, for instance by shortening major revision periods, and at the same time upgrading these facilities so as to increase production.[13] Like in the case of the Millstone facility in the 1990s, this is not without risks, considering the number of unscheduled stoppages at Swedish power plants in recent years. While managers continue to reduce costs, there will be a minor impairment here, another small deterioration there. Meanwhile, it is difficult, if at all possible, to get hold of the general loss of quality, which accumulates over longer periods of time:

> Thus, the short-run savings that accumulate with cutting corners on maintenance and safety can be expected to dominate management thinking at the top, middle, and bottom. Since any untoward consequences of short-run savings are unlikely to appear, if they ever do, until the distant future, management can escape accountability.[14]

Even if deregulation itself will not necessarily lead to the kind of relegation of safety that, arguably, has taken place within large sections of corporations like the government-controlled Swedish power company Vattenfall, particularly up to the incident at Forsmark 1, a facility owned by Vattenfall, in 2006, it seems to have fostered a culture of short-sightedness. In this case there had been insufficient long-term maintenance of facilities, episodes of recruitment freezes, a reduction in the time each operator must spend as a trainee before being allowed into the control room, as well as the overall time spent in training and education. Parts of this may be explained by an increased cash-consciousness of executives and upper management, but also as a result of the ambiguity of political decision making.[15] This is something of a double-edged sword, or *Catch 22*: As the Swedish Government is a major owner, when facilities or individual power plants run into difficulties it is often somehow a result of political ambiguity. Likewise, when facilities show poor results, or other signs of weakness, it soon becomes a source of political ambiguity.

On the whole, many enterprises fail and their strategies turn out to be counter-productive when approaching goals and objectives too directly rather than considering the complexity of "the relationship between the whole and it parts", as well as the impact of the more open process of discovery and adaptation. Planned urban environments like Brasilia and Chandigarh could never imitate, let alone transcend, the vitality or creativity of real communities, by setting out to "create" a vital community. Likewise, successful companies often deteriorate when executives are predominantly looking to "maximise shareholder value". In both cases the interrelationships between high-level objectives and various intermediate goals and actions become too *thin* and simplified.[16]

> The hope that rational design by an omniscient planner could supersede practical knowledge derived from a process of adaptation and discovery swept across many fields in the course of the twentieth century. This approach was generally described as modernism.[17]

Even the progressive utilitarianism of F. W. Taylor was accompanied by a reservation as to the risks of swift and radical change. In his words, these developments ought to be guided by "the right spirit"; the imperatives of scientific management and enforced standardisation should be the greater good of the establishment and, ultimately, the whole of society. In modern society this is not always the case, and short-term gains, profits and efficiency measures are not always beneficial in terms of the long-term vitality and productivity of organisations.[18]

More often than not, corporations that are single-mindedly focused on profits, growth or on being a "top-tier company", are less likely to flourish in the long term. The route to success is often more "indirect"; this is true also with regard to high-risk organisations. Due to a shift in leadership, priorities or corporate culture, organisations sometimes embark on a steady drift towards degeneration: Continuous reductions of costs, by means of downsizing or outsourcing, are likely to have negative effects on the general quality of practice; negligence of experience-based skill and tacit knowledge can have similar impacts. In the longer term such deterioration of quality will most likely lead to some sort of organisational *limbo*, a downward spiral in the process of which previous success formulas fall into oblivion, or to 'an imaginary place for lost or neglected things'. In other words, a process of degeneration.[19] In high-risk activities, this might mean disaster.

4.2   The Measurement of Performance

Various attempts have been made to define the concept of safety culture, and to measure it; to establish certain performance indicators in order to evaluate the safety cultures of power plants worldwide. The aim of developing relevant approaches in order to evaluate the performance of different power plants, as compared to other plants, has been to arrive at objective methods and characteristics, compliant with the entire industry. Yet this can also be misguiding and counter-productive, as key issues of operation are excluded in these measurements, such as the ability to manage the unexpected.[20]

In Sweden, the tendency from the early 1990s onwards has been that all nuclear training and formal education are to be systemised. The main approach utilised to design and evaluate programmes when training nuclear plant personnel is Systematic Approach to Training (SAT), endorsed by both

WANO and IAEA, and by Swedish supervisory authorities. The objective of this Knowledge Management-type strategy, not implemented by TEPCO at Fukushima Daiichi but accepted as international best practice, is to distinguish and break up relevant skills and qualifications; identifying specific tasks to be trained, and to utilise training for performance improvement. Once developed for the US Air Force, so as to speed up and make learning more effective, the SAT method encourages the learning of what you *need to know*, not what is *nice to know*, on the basis of task analysis.[21]

To make training more "efficient", each section of the work is analysed in terms of what distinctive knowledge and attitudes are required to perform a specific task. In distinguishing relevant skills training is required in, through analysis of documents such as instructions and standard procedures, and by means of surveys, observations and Table Top-interviews with theme specialists, form the basis of training. The objective is that each task has a distinct starting point and a clear end, a limited continuance that can be observed, sequenced and measured. Echoing of the ideals of Taylorism, this presumes that each position can be discriminated within the framework of these specific tasks. Besides, in the systemised model practice is not actually a significant part of education; it is the formal aspects of learning that prevail.[22] Adopting methods or strategies from the US Army is usually seen as a sign of credibility and efficiency. However, the US Army does not always succeed, for instance when approaching matters too directly by means of rational estimates and forecasting into the future; if ignorant of the complex environment it faces; or if failing to understand "the means of achieving its own objectives".[23]

As in medical care, it is insisted upon that incidents and deviations from normal procedure shall be reported and documented into an all-embracing enterprise system. The coordination system utilised within Swedish plants, in order to make nuclear power safe, is the Corrective Action Program (CAP). The idea behind this system, promoted by WANO and the Swedish Radiation Safety Authority (SSM), is that, by reporting all sorts of incidents and deviations from "normality", the recurrence of error will be avoided and organisations will share experiences. An ideal condition is defined where everything is considered to be safe, and from which any deviation is reported and registered. From this, new rules and recommendations are issued through the trending of upsets and incidents, and in converting the calculations of risk analysis into observable conditions for safe operation. The function of risk management, therefore, will be to monitor that "the over-all level of safety matches the acceptance criteria".[24]

Still, this is essentially a model of reality, not reality itself. The positives of this may well be to get people to actually report in the first place and, to a larger extent, encourage this kind of endeavour. The negatives, on the other hand, if relying too heavily on a system like this in the calculation of risk might be that, in concentrating on quantification, like the trending of events, upsets and incidents, and assessing the probabilities of future contingencies, does one in fact anticipate and detect the "Black Swan"; that one irregular

event or deviation that might lead to a large-scale accident of massive consequences if not perceived and dealt with sufficiently? In other words, is it a
good model of reality? Does it give a sensible, or a rather askew, depiction of
the actual risks inherent in the system?

> Rare events have a certain property – missed so far at the time of this writ-
> ing. We deal with them using a model, a mathematical contraption that
> takes input parameters and outputs the probability. The more parameter
> uncertainty there is in a model designed to compute probabilities, the
> more small probabilities tend to be underestimated. […] This of course
> explains the error at Fukushima. Similar to Fannie Mae. […] small prob-
> abilities increase in an accelerated manner as one changes the parameter
> that enters the computation.[25]

This is the non-linearity of risk and there also seems to be some sort of
"ingrained tendency in humans to underestimate outliers – or Black Swans".[26]
Many times, the computation of risk, in particular when it comes to small
probabilities, is what produces this false sense of security and understanding.
Besides, within the nuclear industry the insights of safety assessments
have mostly been used to reduce design vulnerabilities that could lead to
beyond-design-basis events rather than focusing on "the mitigation of the
consequences of those events".[27] There might be immanent causes and latent
failures, which can spawn new accidents in various untoward circumstances.
However, upsets, near misses and accidents may not all be triggered by the
same causes.[28] So, we might be heading towards a more comprehensive
account of reality. But then again, is it truthful?

Ultimately, safety responsibilities reside in the companies themselves, while
supervision is handled by national regulatory authorities. I have argued that
the form of auditing processes performed by WANO, an organisation established after the Chernobyl accident in 1986, has contributed to the increased
formalisation within the Nuclear Power Industry, promoting the convergence
to international best practices.[29] Some people have disagreed with me, stating
the fact that the pursuing of WANO guidelines and peer reviews are *voluntary*, not compulsory. When a WANO Peer Review team visited the Ringhals
facility on the west coast of Sweden in April of 2013, typical findings were
deficient behaviours in relation to the use of management tools such as
Self-Regulation, Pre-Job Briefing and Post-Job Debriefing, and in general,
in the control room and around the power plant itself. Furthermore, the
review team emphasised that work methods and standard procedures were
not always consistent and in accordance with industry best practice, as advocated by, for instance, WANO themselves, that is, an international selection
of highly qualified individuals, or *peers*, from all around the world. Even if
some of the developments elaborated on here, such as increased formalisation
of work, are not deliberate effects, but delayed second-order consequences of
such auditing, all Swedish nuclear stations have had the aspiration of reaching

the upper quartile level of WANO safety leading indicators. In recent years though, Swedish facilities have had significantly lower track records than, for instance, American power plants. In fact, Swedish power plants are no longer considered to be among the best facilities in the world. Besides, there has been a tendency at nuclear plants in Sweden that one prepares for inspections and peer reviews rather than for an ambiguous future, in other words for reality.[30]

The KSU has on several occasions been reviewed by WANO, OSART and other external reviewers, answering questions such as: Are routines and processes properly implemented? If models and standards are complied with, then we must surely have quality throughout our organisations? The objectives of these evaluations are to help organisations to improve on the bases of international best practice and operating experience, all of which have favourable aspects to it. To receive various ISO certificates and the thumbs up from external assessors are important sources of legitimacy to many organisations today. The question is: Is it reasonable in a couple weeks, by means of external inquiry, to penetrate the strengths and weaknesses of an entire establishment? Moreover, what delayed consequences or long-term effects will this have on the cultures and qualities of high-risk organisations?[31]

Looking at the case of the higher education, there are similar problems. In the US, the ranking of colleges and universities has had a significant impact on the quality of activities in recent decades. Here, too, these evaluations favour benchmarking and increased harmonisation. Aiming at top rankings, many schools of higher education expand their non-core activities with expensive student centres and excessive administration, sponsored by increased student fees. Ultimately, most students can no longer afford to enter the most prestigious colleges. Besides, there is an urgent risk of decreases in quality, as efforts are directed away from core activities. More often, these developments also generate less rigorous demands on students when it comes to work load as well as intellectual requirements, giving rise to the delusions of a so-called *soft curriculum*. Many colleges do not ask much of students academically, and students are habitually opting out of more challenging subjects such as mathematics and foreign languages. Students have become "consumers" who cherry-pick what they find amusing; they pay good money for it and it is best to keep them happy. Nowadays more people go to college, while the overall failure to graduate has increased. The idea of liberal education and the ideal of *learning how to learn* are for this reason undermined.[32] The obsession with organisational rankings, therefore, is a potential source of degeneration.

The review of which nuclear facilities are the best performers can have similar consequences: organisations such as INPO and WANO carry out inspections and these are combined with the aggregate of measurable indicators – such as the number of unscheduled stoppages or the number of days of unscheduled stoppage – to give an overall performance ranking. In the 1990s Forsmark was hailed as one of the best-performing nuclear facilities in the world, but was later, after the events in 2006, regarded as somewhat of a problem child as some sort of degeneration had occurred over a period of 10 to 15 years. In retrospect, complacency alongside a culture of self-containment have been put forward as contributors to these ill-fated developments, also delineated as a long-term decline in safety culture, mainly due to a lack of a durable strategy of leadership, promoting safety over profit.[33] Another critical factor is the interpretations made by the management of each power station of the secret reports and recommended areas of improvement suggested by the WANO Peer Review team or OSART, promoting a convergence to international best practice. If safety and quality are to be maintained in the long run, plant personnel are in need of appropriate conditions for learning.

Typically, the problem with an obsession with rankings, benchmarking and the measurement of performance often manifests itself in organisations trying to take a too "direct" route to success, albeit externally justified; trading various indirect or *oblique* approaches to quality for more narrowly defined goals and objectives to be fulfilled. An example of this is the Swedish nuclear industry where direct approaches, such as reaching the upper quartile level of WANO safety leading indicators, have successively gained ground at the expense of local knowledge. Likewise the projection of quality can be a threat to the vitality of organisations that are fixated with rankings, quantification and enforced standardisation. Equally, a higher level of proficiency is best achieved indirectly, allowing for deviations from formal practices, discovery and interpretation, in the initiatives of operators "to uncover the moment-by-moment truth of system conditions", identifying both anticipated and unanticipated interactions.[34]

Complex objectives are hard to define and so are the means of achieving them. Policy implementation is seldom clear and simple, and there are limits to our capacity of specifying the potential risks we are dealing with. Moreover, when it comes to high-level objectives such as "world-class safety" or "operational excellence", it seems that employees do not actually know what they stand for or how to translate them into intermediate goals or actions.[35] In Sweden there has been plenty of confusion around such direct conceptions and approaches to safety and quality, and even investigations into their actual significance. This, if anything, tells us something about the dilemmas of being too direct when it comes to issues of quality. It also tells us something of the difficulties of emulating the success formulas of other companies or organisations; the hazard of putting local knowledge in the shade for an alluring projection of quality.

As Michael Power has argued, it is reasonable to make a distinction between auditing and an auditing style of monitoring, oriented towards normative standards, and *inspection*, or an inspective style of management, concentrated on evaluation. Nevertheless, divisions like these are intangible and difficult to differentiate. Usually, auditing processes have a dominant language of evaluation, as well as something of a self-preserving structure. If organisations become more transparent and legible, they will be easier to audit and it will take less time, making it more convenient for all the parties involved.

Eventually it becomes a matter of empirical scrutiny; to what extent these arrangements and activities of inspection and certification have a long-term impact on practice, not only in terms of neutral verification, but as an agent of change, oriented towards compliance and the measurement of perform-ance. In other words, audits are potentially indirect methods of control, "a style of risk processing which in many cases are not neutral with regard to concepts of individual and organisational performance, but shapes them in crucial ways".[36]

As in the case of KSU, there are usually some suggested areas for improve-ment as well as acclaim if a few best practices have been implemented, for instance as regards training and further education. What peer reviewers are mostly looking for is to what extent the uses of operating experiences and best practices are being adhered to; if there are no such spotted "weaknesses" or areas for improvement there is usually a "sigh of relief". But even with the credibility that comes with a positive audit or evaluation, or the embar-rassment of being rejected, legitimacy does not equal quality. It may also inhibit an organisation going forward, addressing those areas where there is genuine need for improvement. Many times, this kind of external legitimacy is also something for executives and upper management to hide behind so as to rationalise activities, pleading for more apprentices for each instructor or by "speeding up" the course of education.[37]

Recently though, WANO have also highlighted global trends and indica-tions pointing at some of the shortcomings of enforced standardisation. In a Significant Operating Experience Report (SOER), a summarising analysis on Significant Event Reports (SERs), "Operator Fundamentals Weaknesses", a critical point of analysis was this: an overreliance on processes and pro-cedures, promoting a compliance-based approach to tasks and a "checklist mentality". Operating Experience in recent years has revealed two recurring problems with regard to the application of standard procedures: (1) opera-tors followed procedures exactly as written, but did not have the in-depth knowledge to know how the plant should respond to a certain situation and condition; and (2) events have occurred when operatives did *not* follow pro-cedures as written, or did not use human performance techniques properly to support procedure use.[38] This may seem contradictory, but corresponds to the empirics on the Swedish nuclear industry, which has been generated in collaboration between the Department of Skill and Technology at the KTH Royal Institute of Technology and the KSU.[39] Here too, many informants bear witness that, following the upgrading of facilities and the display of more "accurate" information in the control room in addition to the fact that the sheer amount of manuals and instructions have increased, there has been a general decline in what is commonly referred to as the "process perception" among plant personnel:

> In recent years many instructors have observed flaws and deficiencies in the process perception of some operatives, and in a number of cases it

has been necessary to go over certain moments of training repeatedly. On some occasions operators have followed specific instructions in the simulator point by point, turned pages at the end, and without thinking gone through an extra page not reacting. In other words, they had not been aware of, and considered, the next step of the sequence.[40]

Apparently, this is not something that happens overnight. As established in the study field of professional skill, rather, over a period of four to five years, one will have a better chance to assess if the skill within a certain organisation progresses or whether it degenerates.[41] More important still than becoming tangled up in detail and measurement data is the formation of an overarching perception and ability of judgement, on the basis of the "fairly accurate" aspects of which simulator training and continual education should be able to reinforce rather than undermine. Attention to correctness and standard procedures is by no means the crucial part in attaining the comprehension of process perception. What is more significant are the various "flows" and patterns in the process; to sense and observe the difficulties encountered, and to detect minor changes.[42] This is what process perception and the skill of anticipation is essentially about, to distinguish the subtleties of change:

> Small events (can) have large consequences. Small discrepancies give off small clues that are hard to spot but easy to treat if they are spotted. When clues become much more visible, they are that much harder to treat. Managing the unexpected often means that people have to make strong responses to weak signals, something that is counterintuitive and not very "heroic".[43]

Small failures can be a plane in the wrong position on a full deck, a sound that is not right or a minor irregularity in the process. Yet in many high-risk activities the slow steady drift towards formalism and automation, obscuring the complexity of systems, as well as the limits of our capacity for abstraction, has induced a shift of attention towards those aspects of quality that are quantifiable, easier to categorise and put a label on. Likewise, things we cannot measure, or manage, have thus become less significant. This also has a strong impact on the direction of training and further education.

There are recently observed effects in aeronautics that the automation of aircrafts was in fact under-challenging pilots, "dulling" their awareness and skills, resulting in more and more accidents. Here too, a Federal Aviation Administration (FAA) regulation has forced the industry to increase its level of automation in the promotion of safety. Ultimately, supervisory authorities realised that the reliance on automated flying was defective and that pilots often abdicated too much responsibility to these automated systems.[44] As in the case of the disaster with Air France flight 447 from Rio de Janeiro bound for Paris on June 2009, the high level of automation and the modern design through which pilots control this aircraft in a highly abstracted manner was intended to reduce both risk and the workload of pilots. Yet in this case the reduction of pilot influence, in the spirit of computerised flight control and "enforced non-involvement", made pilots unaware of what was really happening, as they unknowingly stalled from high altitude all the way down into the ocean. Once the digital flight-control system failed, the pilots did not achieve the information needed to detect what was going on; neither had they received sufficient training, as this was something that was never supposed to happen. Moreover, the design of the system, while intended to moderate the potential for human error, implied that they were also less able to uncover from unsafe situations when faced with the unexpected:

> The problem is that while accomplishing this simplified and clear-cut work environment, the design separates also the pilots mentally and emotionally from the complex reality within the atmosphere, and from the basic dynamics of flying. The pilots are forced to pay most of their attention to detached knowing-that the plane will follow certain high-level orders, instead of being continuously involved with knowing-how to make the plane do what is desired.[45]

In other words, people often like to think that technology has more intelligence that it actually does. As we are afraid of minor upsets and variability, we are continuously looking to create "error-proof" systems, overprotected by formalism and technology, notwithstanding the fact that, typically, as in the case with the Air France flight 447 from Paris, this "avoidance of small mistakes makes the larger ones more severe".[46] By stating this I am by no means questioning the so-called defence-in-depth principle of nuclear stations, which is of course an integral part of the safety of these technological systems. It is rather that the urge to marginalise the manual skill of operatives can be associated with the overprotection of *any* system and the long-term risks that come with it. This is also something to consider in nuclear operations. In recent decades, it seems that each added measure of technological fixes are believed to enhance the defence-in-depth principle. Up to a certain point this may well be true, but it may also become counter-productive, adding to the complexity of these systems and the various "opaque causal links" of potential interactions.

Örjan Ekwall, Operational Manager at Forsmark, argues that the pursuit of formalisation within the Swedish nuclear industry in recent years, promoting detailed rules and regulations rather than *frameworks*, has entangled its different partners into a structure that is now difficult to find a way out of. There has been ambiguity around, for instance, situations where operatives have taken the right steps or measures, but not followed a given instruction. Though it is the symbiosis of instructions and operator know-how that is assumed to assure quality, to make operations safe and reliable, in circumstances where uncertainty is ruled out, the risk of stagnation is immanent:

> What is our reaction when something goes wrong? We have an inclination to blame it on an ambiguous instruction, which we then immediately make more precise. And so we keep going until we have a mountain of instructions, so detailed that there is little room left for manoeuvre [...] We have a complex process to master, based on thousands of instalments that can vary with regard to each other continuously. This implies that a given situation is never quite the same. For that reason, it is important for operatives and plant personnel in all areas of the facility to sense this process, and to observe minor deviations by means of process perception. It is impossible to consciously process all this information in real-time situations of practice. Rather, the information forms a pattern that operatives must learn to recognize and interpret. This imperative know-how cannot be captured by any instruction, neither taught in training. The only way to get hold of such knowledge is to reside in these specific environments, to actually manage the process over longer periods of time.[47]

In other words, rules and instructions are never actually followed to the letter. Consistent with Örjan, we need to add one aspect to the conception of quality that is predominate throughout this sector of industry, the kind of personal judgement and situation awareness that cannot be measured and formalised for the benefit of learning and formal education. In parallel with this enforced standardisation, within these operating organisations there is also an inherent structure of recruiting and promoting plant personnel, where each new employee has the same basic job training and education. Those who prove themselves creditable for promotion will gradually undertake further education to become operators, and at some point later even supervisors, which generally takes about 12 to 15 years.

It generally takes a combination of training and practice to attain these positions but it is the experience-based proficiency that is the most decisive. However, in view of the systemised model, practice is not an intrinsic part of the overall education. Consistent with this approach, formal education is regarded as the single most important aspect of professional knowledge and learning, for instance as regards the qualifications required for promotion. The aspiration to formalise qualities that are elusive of detailed description is primarily a concern to people outside the context of practice, whose job it is to assess how well the system works compared to normative standards.[48] The experience-distant rationality of auditing and risk management is in pursuit of control, taxonomy and systematic understanding. In other words, people distant to the actual situation of practice need fixed standards to work with, even though this has also created a cycle in which operating organisations are pushed in a direction towards enforced standardisation. In the course of these long-term cultural changes of work, hence, the criteria for "good work" have become more and more formal.

In other words, skill is not so much about the management of technology. It has more to do with the formation of considered judgements, on the basis of analogical thinking. Many of the informants that participated in the above-mentioned study point to the fact that operatives nowadays are clever at detail but often seem to lack the overall picture and perception. Where formalisation is stretched, there are reasons to suspect that it proceeds at the expense of the resilience of human professionals and their "subtleties of application". This is the ironic nature of the kind of designs for work and production that tend to diminish the agility and initiative of their intended beneficiaries.[49] This approach may be regarded as too informal when it comes to high-risk industries but these features of quality, based on learning beyond the context of formal education, are in fact vital to the maintenance of a dynamic safety culture, in analogy to the accumulation of skill and manual knowledge in other fields of proficiency:

> Sometimes it's imagined that becoming skilled means finding the one right way to execute a task, that there is a one-to-one match between means and ends. A fuller path of development involves learning to address the same problem in different ways. The full quiver of techniques enables mastery of complex problems; only rarely does one single right way serve all purposes. The rhythm of building up skill can take a long time to produce results.[50]

Within the Nuclear Power Industry the idea is that the effects of training are to be measured, as a way to assure its quality. Testing of short-term memory will measure the outcome of training and formal education, for instance if attending a course in how to carry out a new routine or best practice. The actual effect, however, whether key members of staff have become more enlightened or skilled, will always be difficult to measure. A lot of the learning that goes into skill is informal. For that reason it is often rather unconscious, attached to specific environments and situations, to some extent "hidden in action".[51]

The fact that international organisations like WANO have begun to highlight the risks with regard to checklist mentalities, along with compliance-based safety cultures, and that they now seem to have de-emphasised the concept of Knowledge Management, may indicate some sort of progression of the organisation as such alongside its approaches to learning and issues of quality. In Sweden there has also been an inclination in recent years to decrease the number of routines and instructions rather than the opposite. The question is: What is the key to a dynamic safety culture, and what approaches to supplementary training and education can be utilised to support the reflective processes of experience-based skill and knowledge?

Notes

1 Spelplats 3.2009, p. 60; Spelplats 3.2010, p. 66.
2 Woods, D., Dekker, S., Cook, R., Johannesen, L. and Sarter, N. (2010): p. 8.
3 Ibid., p. 118.
4 Perrow, C. (1999): p. 11.
5 Ibid., p. 79. Cf. Woods, D., Dekker, S., Cook, R., Johannesen, L. and Sarter, N. (2010): pp. 141f.
6 Perrow, C. (1999): pp. 84–85.
7 Perrow, C. (2011): *The Next Catastrophe*, pp. 9–10. We can here talk about different *types* of failure, like that of operatives, that is, human error, management or executive failure, the complex interaction of the system accident, or even "ideological failure" of, for instance, deregulation; demarcations that can be useful for analytical purposes. While the Forsmark accident was a typical case of system accident, the accident at Fukushima Daiici could well be classified as an executive failure, given the severe critique of TEPCO put forward in the NAIIC report.
8 Rasmussen, J. and Svedung, I. (2007): p. 14.
9 Göranzon, B. in Göranzon, B., Hammarén, M. and Ennals, R. Eds. (2006): "Tacit Knowledge and Risks", p. 196.
10 Perin, C. (2007): pp. 3–4; Corradini, M. and Klein, D. (2012): p. 28.
11 Perrow, C. (2011): pp. 155–156.
12 Ibid., pp. 157–164.
13 Berglund, J. (2012): *Säkerhet och ekonomisk rationalisering* (Safety and Economic Rationalisation), pp. 31f. The report was written on behalf of the Swedish Nuclear Safety and Training Centre (KSU).
14 Perrow, C. (2011): p. 144.
15 Cf. Larsson, L. and von Bonsdorff, M. (2007): p. 14. In the case of Vattenfall, cutting costs on maintenance and operational safety can also be linked to dubious company take-overs abroad, like in the notorious case of Nuon Energy. Conversely, as argued by Ferguson, N. (2014): pp. 58f, the dispute on deregulation is sometimes based on a flawed understanding of how markets actually work. In other words, it is not deregulation itself that causes these problems, but rather a lack of individual and corporate responsibility. In many cases, such as the financial crisis of 2009, the prime evil is rather the excessive complexity of regulation, described as "the disease of which it pretends to be the cure".
16 Kay, J. (2011): pp. 50–55.
17 Ibid., p. 4.
18 Taylor, F. W. (2007): pp. 113–115. Cf. Sennett, R. (2006): p. 168f. As Nordenstam, T. (2009): *The Power of Example*, pp. 201–202 has clarified, the principle of utility "neither can nor need to be proved", and in the case of Taylor the maximisation of long-term happiness, or utility, for all parties involved could be achieved only through maximal prosperity and productivity. In this sense, Taylor worked in the spirit of Enlightenment, while inevitably touching on some of the challenges and perplexities of modern-day capitalism.
19 Berglund, J. (2013): pp. 184f; Perin, C. (2007): pp. 257–267; Scott, J. C. (1998): pp. 350f; Kay, J. (2011): pp. 20–24. Scott, J. C. (1998): p. 356, points to a similar paradoxical texture with regard to the kind of designs for life and production that tend to diminish the skills, agility, initiative and morale of their intended beneficiaries. The concept used by him to characterise the corrosion and undermining of local, experience-based knowledge, what he refers to as Métis, is "institutional neurosis".
20 Perin, C. (2007): p. 15.
21 The uses of the SAT method within the Nuclear Power Industry are discussed in Berglund, J. (2013): pp. 91–96.

22 Spelplats 4.2014, pp. 7–8.
23 Kay, J. (2011): pp. 92–96.
24 Rasmussen, J. and Svedung, I. (2007): p. 42.
25 Taleb, N. (2013): p. 454.
26 Taleb, N. (2010): pp. 141f.
27 Corradini, M. and Klein, D. (2012): p. 23.
28 Cf. Rasmussen, J. and Svedung, I. (2007): pp. 27f.
29 Cf. Berglund, J. (2013): pp. 87f.
30 Cf. Spelplats 4.2014, pp. 17–19; 33–35.
31 Spelplats 2.2014, p. 35.
32 Cf. Murray, C. (2008): *Real Education – Four Simple Truths for Bringing America's Schools Back to Reality*, pp. 96–116f.
33 Larsson, L. and von Bonsdorff, M. (2007): p. 10.
34 Perin, C. (2007): pp. 213–214.
35 Cf. Kay, J. (2011): pp. 87f.
36 Power, M. (1997): p. 140.
37 Spelplats 2.2014, pp. 36–37.
38 Interview with Hans Ehdwall, Senior Advisor at the Swedish Nuclear Safety and Training Centre (KSU) and a former WANO Interfacing Officer (WIO), Stockholm, 18 October 2013.
39 A large part of this empirical material is documented in the periodical Spelplats, see Literature. I have used parts of this material in Berglund, J. (2013). For this study I have looked into the new empirics being generated through this collaboration and I have also given some of the old material a "second look".
40 Mats Brändström, Head of Training KSU Oskarshamn, Spelplats 1.2011, p. 30. My translation.
41 See Göranzon, B. (2009) [1990]: pp. 126f.
42 Mats, Head of Training KSU Oskarshamn, Spelplats 4.2014, pp. 15–16.
43 Weick, K. and Sutcliffe, K. (2007): p. 8.
44 Taleb, N. (2013): p. 43.
45 Stensson, P. (2014): *The Quest for Edge Awareness – Lessons Not Yet Learned*, p. 25.
46 Taleb, N. (2013): p. 85.
47 Örjan Ekwall, writings prior to Dialogue Seminar on Safety Culture, 4 November 2010, at the KTH Royal Institute of Technology, Stockholm. My translation.
48 Spelplats 4.2014, pp. 7–9.
49 Scott, J. C. (1998): p. 316.
50 Sennett, R. (2012): *Together – The Rituals, Pleasures and Politics of Cooperation*, p. 201.
51 Gustafsson, L. and Mouwitz, L. (2008): pp. 17–18.

5 The Skill Factor

5.1 Dynamic Safety Cultures

The case study introduced in Chapter 4 on nuclear education and supplementary training points to the fact that the style and focus of training has shifted, from process perception to the direction of conceptual knowledge and enforced standardisation. In that sense it has been convenient to observe false behaviour rather than to underpin the experience-based skill of plant personnel. Certainly, setting up various efficiency targets can motivate higher costs for training. This means there might be a point in underlining where the conceptual knowledge is obtained, where skill and knowledge of familiarity originate from and who is responsible for which.[1] It appears that management sciences like Taylorism and Knowledge Management tend to favour and give credit to the former, those aspects of work-related knowledge that can be, or already have been, formalised.

Nowadays, there are fewer faults and deviations at most power plants built in the 1970s and 1980s; initially the untried plants had plenty of minor upsets and deficiencies. Likewise, there has been a decrease in the learning obtained from everyday practice and experience, partly due to increased automation, and hence fewer payoffs from action, good or bad. Besides, there is also an ageing problem ahead, while decisions have been made to extend the life of power plants originally designed to run for 20 to 30 years, for instance due to the fact that certain resolutions to abolish nuclear power in countries like Sweden have been withdrawn:

> Our database of accidents cannot tell us how serious the aging might be. We don't even have very good evidence as to whether the reliability of our plants is stable, going up, or going down. Numerous measures are available; but they point in different directions and are hard to interpret. Perhaps the most clear and simple one is the number of serious incidents (i.e. near misses). Here the news is mixed. The number dropped from 0, 32 per reactor year in 1988 to a low of 0, 04 in 1997, perhaps reflecting a maturing of the industry and few plants coming online. But it then rose to 0, 213 in 2001 – about twenty-one serious

incidents a year if there are one hundred plants operating all year – perhaps reflecting the aging problem.[2]

These numbers reflect the operation of US power plants, which in an international perspective are estimated to be at the top end in terms of safety and reliability, but this evolutionary curve corresponds to the situation at Swedish power plants where availability numbers have fallen considerably over the past decade. In other words, we have plants designed for 20 to 30 years of operation that are looking to extend their utility by another 20 years. The increase in number of incidents might also reflect the long-term effects of a reduction in maintenance costs, and operating costs in general. If maintenance activities are neglected it will add to the complexity of the facilities and increase the potential for unanticipated interactions and system accidents.

In a Swedish context there is also the issue of an ageing workforce, the challenges of an ongoing generation shift in terms of plant personnel and other members of staff. Consequently, recurrent training in full-scale simulators has become more accentuated, practising courses of events and disturbances that nowadays rarely occur. Stressing the importance of experience-based skill and learning does not mean there is no need for administrative systems such as support, keeping regulations, instructions and information up-to-date.[3]

Employees that are recruited today are not of the same background as during the build-up period of nuclear power in countries like Sweden, when, most notably, people with experience from machine work at sea and from the Marines were employed. Apart from the fact that these individuals had technical skills, they brought in an affirmative approach to learning, and more specifically the importance of learning from everyday practice, and were able to share their knowledge with others. As in nuclear industry, in the Marines you train to prepare for things that are hopefully never going to happen.[4] In other words, these people brought with them a positive cultural impact on learning and a cultural imperative for cooperation.

As a result of the lack of learning from everyday practice, "hands-on" recurrent training in full-scale simulators has become more accentuated. With this kind of training, it is often assumed, operatives are able to prepare for a real-time course of events, allowing them to calm down in stressful situations: Training in full-scale simulators is an important preparation for many high-risk activities; it can help us to "cool down" when being challenged in situations of crisis, by trying to enhance the kind of readily available, tacit knowledge vital when the pressure arises:

> In such situations we are forced to trust previously learned strategies in dealing with serious problems. In aviation, the pilots spend part of their training in simulators in order to learn "by heart" items, which are crucial in critical operations. Operators in other safety-critical systems like nuclear power plants have the same kind of simulator training.[5]

In my enquiries into the annual simulator training programme for operatives at Forsmark 3, the main target was to make all shift teams perform and cooperate in the same manner. The task was for each group to detect unacceptable Core Oscillations in the reactor in a simulated scenario. As it turned out, much to the dissatisfaction of the instructors, each of the six shift teams unravelled the scenario differently. The instructors would have preferred all teams to act in the same way, namely to identify the unwanted oscillations in temperature by means of a small technological device, the more recently implemented 520-alarm; the indication being that the colour of its displayed measured value shifted from green to yellow.[6]

The shift from analogue to digital technology has given rise to uncertainties that are now being addressed in training and further education. Arguably, the shift to digital screens has also reduced the possibilities for operators to read patterns, and by means of sensory experience to learn how to detect minor changes in the process. Nowadays there are fewer patterns to read. In other words, the murkiness of abstraction and fancy graphical representations is "a potential loss of detailed insight".[7] The increased propensity of Core Oscillations detected in recent years, as a result of modifications to the composition of fuel, was highlighted by the implementation of new technology for this particular upset, in this case the 520-alarm. While this is approached in training as adjustments being made to one "block" of knowledge or another, we will not know what comes next, or what is really that "Black Swan" – rare events of vast consequences which may lead to a severe accident. If a certain risk occurs only rarely we tend to underestimate its likelihood. But if something *does* happen, such as the terrorist attacks of 9/11, or the accident at Fukushima Daiichi, we then tend to exaggerate the risk of similar events occurring in the near future.[8]

There is also the risk that formal educational qualifications do not cover the actual needs with regards to knowledge. When discussing issues of learning and recurrent training, educators have observed that, in the simulator, operatives tend to solve problems or situations in different ways. As in the case of the Forsmark facility, the units of Forsmark 1, 2 and 3 have different cultures and praxis that have emerged over time. This is partly due to tradition and the principles of leadership that prevailed, as well as the fact that there was competition for many decades between the three blocks. These differences can also be spotted in training. In the simulator, operatives by and large reach the same targets yet do so differently, using slightly different approaches. For many years there was little or no cooperation between the blocks. Nowadays, on the contrary, one has to struggle to pursue something of one's own.[9]

Full-scale simulators are utilised in the training of operatives in many high-risk industries. Still, there are many differences between various practices. Some groups, like physicians, are not used to feedback, and in, for instance, the shipping industry, it is unusual that people are called back for any sort of requalification, unlike in the nuclear industry or in aviation. The increased significance of supplementary training has highlighted the role of good instructors' skills, which have also become more intrinsically associated with formal education. In Sweden, pedagogical approaches have been largely didactic, with an emphasis on tests of short-term memory. Experienced operatives are typically regarded as not being flexible enough, for instance in adapting to new technology; not eager enough to learn, to adopt new methods and best practices. Likewise, scenarios are generally goal-oriented from trends and analysis of Operational Experience, with instructors anxious not to promote "incorrect behaviour". In fact, some would even argue that the overall prospect has been one of control, to monitor people in such a way that the impact of the irrational human factor is minimised.[10] Arguably, this was not how it was originally meant to be. Such divisions of credibility are echoes of the long-term cultural changes within this sector of industry, and a slow steady movement towards increased formalisation.

Developments like these are also informed by directives from the supervisory authorities, whose recent scheme has been to promote increased legibility and more "efficient" methods of education. Essentially, they want their dependent organisations to measure that activities, such as quality work or training, will have the same effects as originally intended. Preferably, they want to be able to single out immediate effects of, first and foremost, various educational interventions. At best, the effects of such training activities are assumed to have an impact the very next day, illustrating a shift towards a "transfer of knowledge"-type educational system.[11] Such developments illuminate the focus on legibility and short-term incentives when it comes to learning and formal education.

In the recurrent training of plant personnel, likewise, the pursuit of formalisation is manifested. Trying to claim all the right answers, many instructors get stuck in error prevention. Enhancing the awareness of instructors, for instance on the division between theory and practice, becomes a key issue. Dieckmann et al. (2009), in discussing medical simulation, emphasise the importance of debriefing, as well acknowledging that simulations are not reality; even if the authentic aspect of simulations are often underlined, as in the Nuclear Power Industry, there are limits as well as opportunities to this educational tool. On the contrary, to depart from realism, and the fact that simulator settings are *not* real practice, can in itself favour learning, in effectively utilising the *as-if* character of simulations. After a session, discussions can take place more openly, for instance around what the participants have learned through their very *participation* in these scenarios. Participants might perform actions that are not anticipated during the design of a scenario; the ideal is not to mimic reality but to create a safe educational environment for learning and self-reflection. Reflecting on the collective experiences made during the scenario becomes vital in order to promote learning, while the role of instructors will be to give feedback and ask questions that can trigger reflection, but also to involve those "silent participants" who are less active during the scenario.[12]

Arguably, there is also a difference when it comes to learning whether the participants know or do not know what will happen during these sessions. In nuclear simulations it is fairly easy to design scenarios where the operators will end up in situations where the mere following of instructions and procedures will be misguiding. Occasionally, participants in these training activities, overseen by the instructors, might find themselves in positions or situations that are legitimate in accordance with one instruction, but illegitimate according to another. This is not to depart from reality; it rather reflects some of the inevitable limits of formalisation and rule-guided *knowing that*.[13]

The complexity of practice implicates that attempts of comprehensive codification or regimentation are no guarantee for enhanced safety and reliability. Such approaches may rather make it easier to distinguish human error in terms of failures to follow rules but also tend to generate more fragile work systems. Hence, the dilemma of contradictory rules and instructions, and what it means to follow procedures in complex situations, is nothing new. This has also led to goal conflicts in situations in which operatives are considered in the wrong if they follow procedures that were inadequate, just as they can also be reprimanded for violating procedures that turn out to be insufficient:

> One consequence of the Three Mile Island nuclear reactor accident was a push by the Nuclear Regulatory Commission for utility companies to develop more detailed and comprehensive work procedures and to ensure that operators followed these procedures exactly. This policy appeared to be a reasonable approach to increase safety. However, for the people at the sharp end of the system who actually did things, strictly following procedures posed great difficulties. The procedures were inevitably incomplete, and sometimes contradictory. [...] Then too, novel circumstances arose that were not anticipated in the work procedures.[14]

In other words, simulations are not always the answer. It is difficult to simulate everything that goes on simultaneously in real-time situations of practice. Rules, instructions and standard procedures are merely "models" of the work content, which are bound to leave many degrees of freedom when it comes to decision making, as these descriptions cannot foresee all local contingencies of a future work context. In general, specific rules are designed one by one for a particular task, whereas in concrete situations of practice several "tasks" are active at the same time.[15] Within the maritime industry, one activity that has proved particularly difficult to simulate is the mooring of larger ships and vessels. It contains elements of difficulty that simply cannot be grasped in simulations. Thus, the only way to get hold of this knowledge is to dwell in these specific environments, to sense and observe.[16]

Clearly, design and application of technology are two fundamentally different dimensions of operational safety; during operation rules and instructions are never followed to the letter. By contrast, the designer needs information on the capacity of the system, not about changes and disturbances. There are different opinions within high-risk activities, however, regarding the level of knowledge necessary of the operative, or the pilot and so on. In other words, what technical and theoretical knowledge is required for the operator relative to the knowledge of the designer in order to prevent accidental side effects?[17] Now, we need technology and technological systems to create as well as control nuclear reactions, and we rely on arrays and feedback given out by the systems so as to make judgement and for selecting potential control inputs. In other words, "the human skill factor" is indispensable in order to manage complexity, to protect and monitor the overall safety and continuity of the process:

> The domain of effects (for the plant in general, but particularly for the core nuclear reaction process) is completely separated from the work-domain, especially from the perspective of control room operators. The central problem with all activities conducted from a distance is that available feedback can only be what is technologically feasible to convey and what in advance is assessed as enough valuable to convey that it merits costly implementation of instrumentation systems. This means that for any kind of unpredicted event there might be relevant observations for system operators that the technology fails to convey, simply because such an event had never been thought of before.[18]

Again, professional skill within this field of activities is not primarily about the management of technology. If there is something like an element of creativity within safety-critical activities it has to do with the nurturing of a questioning attitude, critical thinking or even mild paranoia, as regards safety and the anticipation of hazard. When Winston Churchill discovered that the Japanese Army had taken the unexpected route through the Malayan jungle to capture Singapore during the Second World War, he blamed himself and his advisors for lack of imagination: "I ought to have known. My advisors ought to have known and I ought to have been told, and I ought to have asked."[19] While our knowledge of the world is often incomplete, the skill of anticipation also involves many unplanned activities, as well as the continual expectation of developments that do not materialise. These are elements that have to be nourished, or at least not repressed, along with a positive cultural impact, on learning and personal responsibility. In other words, within a certain work culture or praxis, there are indirect sources of quality that will have to be retained.

From the likes of WANO there have been numerous efforts to promote such attitudes, and also to *standardise* the deliberative reasoning of action by means of training, with Human Performance Tools such as Pre-Job Briefing and STARC (Stop, Think, Act, Review and Communicate); a method for *self-control* established as one of the kingpins of reliable safety cultures. Nowadays there are training courses for everything, a new rule for every mistake and, ultimately, added regulations and instructions become "the answer to everything". Experienced operatives, on the other hand, are sometimes reluctant to undertake certain education. They would rather work than to attend courses in "how to take an initiative" or "how to write an instruction".[20]

Besides, the concept of STARC implicates that *time* is the only important factor in good decision making; to make a pause and to think more, at all times. In that way, these tools and methods often become too simplified or superficial. Human Performance Tools such as these are regularly put forward as measures or solutions rather than as support, reflecting also the cultural changes of attitude towards knowledge expressed in practice. In commercial aviation, where airlines constantly struggle with the pressure to depart on time, particularly at airports with high traffic and in high workload situations, the constant push for efficiency has led to numerous problems. Pre-take off checklists have been regarded as a key measure in the creation of safety. Still, this is no guarantee for the counter-balancing of conflicting goals or objectives:

> Leaving both pressures in place (a push for greater efficiency and a safety campaign pressing in the opposite direction) does little to help operational people (pilots in the case above) cope with the actual dilemma at the boundary. Also, a reminder to try harder and watch out better, particularly during times of high workload, is a poor substitute for actually developing skills to cope at the boundary. Raising awareness, however, can be meaningful in the absence of other possibilities for safety intervention, even if the effects of such campaigns tend to wear off quickly.[21]

The message should not be misinterpreted: Formalisation of work by means of performance tools and standard procedures, or by guiding the direction of training, is not unfavourable in itself. It is overdoing it that will pose the greater risk in terms of long-term consequences or side effects; a kind of "if it isn't broken, fix it" approach to quality and human ability.[22] This is not so much being proactive as being short-sighted.

Greater safety returns can only be expected if considering various background factors and larger conditions that may have an impact on other issues of significance, latent contingencies or cultural changes of work. Raising awareness at all levels of the potential movement towards accident and deteriorations of safety is more likely to be more successful than the quick fix approach of telling front-line operatives to "think more" or to "be more careful".[23]

Although a quality system, or formalisation in general, is important to the organisation as such, it has also meant that their practice has had to be accompanied by duties such as the management of documentation, and few people involved seem to have an overall grip on such documentation. Also, most tools and manuals within the area of safety culture provided by WANO that promote, for instance, a "questioning attitude", do not take cultural differences into account. These are based on standardisation; that everyone should act, think and respond in the exact same manner. They rather take for granted that people around the world are equal, however in countries like Japan the challenge of authority is virtually unfeasible. Even if they have demonstrated resilience in dealing with natural disasters in the past, the Japanese desire for order also seems to have had an adverse impact on their emergency preparedness:

> Japan, too, routinely exercises its nuclear emergency management system and has modeled much of its system after that of the United States. But, unlike the United States, Japan rarely tests the limits of the system and training of personnel by using highly unusual events or crafting scenarios that are impossible to recover from. Culturally, the Japanese do not accept failure as a learning opportunity. The Japanese system is largely designed to test the proficiency of the operators in responding to known scenarios. The problem with this approach is that if a scenario has not been incorporated into the design basis, the ability to anticipate and respond is lessened.[24]

Guidelines and instructions will of course make valid sources of information accessible to provide a basis for decision making. Yet, as manual knowledge is replaced by formalisation, functional autism and the aversion to ambiguity have spread throughout this sector of industry. As reality is categorised in terms of black and white, right and wrong, true and false, the reduction of thinking is imminent; merely looking for the one right way to execute each task, influencing the course of training and further education, is not reinforcing the capacity of managing the unexpected. As people of other backgrounds have become more involved in risk analysis, added root causes are taken into account, sometimes to the frustration of more technically oriented colleagues.[25]

Nonetheless, trying to delimit the human impact on safety and quality, and on the facilities, may well contribute to the hollowing out of skill and ability, the visualisation of complexity in dealing with difficulty, and ultimately the ability to intervene in order to manage the unexpected.

In this way the preconditions for building up skills are changing from problem solving – learning to address the problem in different ways, an ability that can be continuously developed – to the kind of rule-following predominantly concerned with find that "one-to-one match" between means and ends. In the long run, leaving behind the kind of learning that comes from analogical thinking and the personal analysis of mistakes, creating a basis for experience-based knowledge, is also likely to alter or rework the character of skill itself. In long-term practices, the learning that goes into skill is evolved through a dialectic interplay between normative rules and standards, and *optionality*; an inclination towards (rational forms of) trial and error, doubt and discovery. If novices are merely presented with "the right way" forward, such progression is likely to struggle with a false sense of security and understanding.[26] The same is true in medical care, for instance when comparing the clinical skills of medical students and doctors:

> The experienced doctor, as one would expect, is a more accurate diagnostician. This is due in large part to the fact that he or she tends to be more open to oddity and particularity in patients, whereas the medical student is more likely to be a formalist, working by the book, rather rigidly applying general rules to particular cases. Moreover, the experienced doctor thinks in larger units of time, not just backward to cases in the past but, more interestingly, forward, trying to see into the patient's indeterminate future.[27]

Conversely, experienced employees may lose the hunger for trial and error if wrapped up in a system of enforced standardisation. If considering that more people seem to get injured at zebra-crossings than haphazardly crossing the street, we seem to benefit from at least some measure of volatility, to heighten our situation awareness. Likewise, looking to comprehend how things react to disturbances might force deviations from standard procedures, exploring the boundaries of established practice.[28]

By continually reducing minor errors and stressors, sources of learning and "antifragility", we are in fact making these socio-technological systems more vulnerable. In this respect, minor errors are often benign and reversible. Besides, they are often rich in "information", for instance on how a certain system behaves, which we can use as "tools of discovery" in the accumulation of skill. We can also get vital information from the mistakes of others and even more so if nurturing a mistake within a praxis. Whereas chronic low levels of stress can be harmful, humans tend to do better with at least some amount of volatility or variability, up to a certain level of pressure. We also need time for recovery and reflection to restore the balance.[29]

> When you are fragile, you depend on things following the exact planned course, with as little deviation as possible – for deviations are more harmful than helpful. […] Further, the random element in trial and error is not quite random, if it is carried out rationally, using error as a source of information. If every trial provides you with information about what *does* not work, you start zooming in on a solution – so every attempt becomes more valuable, more like an expense than an error.[30]

Predictive systems paradoxically cause fragility in the longer run, as we tend to make systems more vulnerable when overprotecting or over-stabilising them, out of our desire for order and our aversion to volatility and ambiguity, which have been the schemes of modernity. This applies to the socio-technological systems of industrial power plants, financial systems, as well as the natural systems of the human body. If a number of artisans, small entrepreneurs or businesses go bankrupt, it is a small price compared to the collapse of an entire socio-economic system. If trying to eliminate all minor upsets and disturbances in order to make systems highly predictive through the systematic removal of variability, there are no stressors to learn from, to brush up our senses and ingenuity.[31]

The overprotection of systems tends to produce small gains, like increased profits or the production of comfort and credibility in the short term but large losses in the longer term, such as added complexity or even the hollowing out of ability, *de-skilling*. Once something happens it is therefore likely to be some rare event of massive consequences, and those overseeing the system are likely to be caught off guard. One dilemma, as within nuclear industry where there is a strong reliance on so-called "stress testing" of, for instance, a reactor, is that risk managers are usually using information from prior accidents as worst-case scenarios, which are then used to estimate future risks as the worst future outcome. But the comfort, or overconfidence, that comes with this type of estimation is bound to overlook the irregularity of the fact that these scenarios, which are currently the worst conceivable, when it happened exceeded (what was considered to be) the worst case at the time. In other words, they were without precedent and thus escaped both expectations and standard procedures. If the *Titanic* had not gone under we would probably have kept building larger and larger ocean liners, and the Fukushima Daiichi accident will most certainly prevent future accidents from becoming even worse. The nuclear industry has become aware of the problems with reactors and flooding, as well as the risks of small probabilities. Hopefully, we have also become sensitive to the delusions of naive stress testing, looking into the future "by naive projection of the past", arranging for the *expected unexpected* rather than preparing for the next future surprise or crisis, anticipating the changing potential for failure. Naive rationalism, for that reason, indicates a single-minded focus on the known, ignoring the *unknown*. Refined rationalism, on the other hand, implies an insight into the fact that beliefs and actions can be legitimate and rational, even if "one does not have the full story".[32]

Within many sectors of industry, developments in recent decades seem to have weakened the prospects of a more dialogical or *open* form of cooperation. In the Nuclear Power Industry the pursuit of formalisation, such as the adaptation to quantifiable targets of quality, safety and effectiveness, has created a growing gap between resource management and practitioners, with less and less dialogue. Nowadays not even a CEO seems to be able to exert that much influence. There used to be substantial dialogue on matters of implication; people actually met, engaged in conversation, got to know each other, and exchanged ideas and experiences. Yet dialogue has to a large extent been replaced by the "fetish of assertion", communication in which "content is all that counts".[33] For the most part cooperation is restricted to forwarding documents and regulations between departments. What has become more important is displaying a confidence that everything is coped with and that all risks are predictable, the same fallacy that has plagued the Japanese. However, as we keep on projecting into the future, scientific-like knowledge has "the remarkable power of producing confidence".[34]

In keeping with Taleb, never have "people who talk and don't do" been more prominent and played a larger role than in our time, the ones in society and in working life who make the predictions of the future, or the "*postdictions*" of the past. People who thought they knew more about the world then they actually did, and let this knowledge guide their decisions, have often created major damage. As the act of decision making is detached from practical knowledge, managers and other professionals tend to safeguard their positions by claiming protection from science and scientific-like knowledge. Also, often we are psychologically vulnerable to the calculation of risk and forecasting, to persuasive narratives and how things are framed. As others are exposed to risks, fragility is transferred to their intended beneficiaries; those citizens, small savers or employees "who suffer society's ingratitude" while others rise in status.[35] This is what is also happening in high-risk organisations. In many areas of work we are inundated with information of all sorts. But this information is not of much worth if we are not equipped to think for ourselves, to draw out its implications and significances. This is the critical thinking aspect of professional skill; to work out for ourselves "what we ought and ought not to believe". We might be persuaded to agree on some conclusion or "truths" which we might otherwise not accept. In other words, the critical thinking aspect of skill can protect professionals from what others seek to make us believe, to be placed in the hands of "the ideas and values of others".[36]

Analogical thinking as a foundation of operators' proficiency cultivates the formation of considered judgements, the refined ability to read patterns, and to see the relevance in situations that are similar but *non-identical*. When we encounter something out of the ordinary, something we do not fully understand, it is a natural reaction to handle it "by reference to something that is familiar to us".[37] The upgraded technology is not predictable in comparison with the old one, and there are a number of combinations that are difficult to foresee. When in the control room, the process is being presented in a new and "improved" manner, but this is no guarantee that the overall safety of operations will increase. The amount of signals, and the information that plant personnel have access to today, does not necessarily enhance the ability of operators to estimate the "well-being" of the process.[38]

In conclusion, owing to the growth of scientific knowledge, the promises of forecasting and the reaches of scientific achievement, we tend to overestimate our ability to understand the "subtle changes that constitute the world". As a result we do not properly discriminate the limits of the calculating logics of engineers and risk assessment, or the type of knowledge we use to project into the future.[39] In other words, in protecting the safety and continuity of production, no factory, power plant or work place organisation can sustain effectiveness in the long run without the unplanned interventions and social interactions of a skilled workforce. Arguably, it is the quality of these interventions that is key to the establishment of a dynamic safety culture.

5.2   The Learning That Goes into Skill

Jean Lave and Etienne Wenger argue that human learning is essentially a process of participation in various communities of practice, within which learning is mainly conceived as an "integral and inseparable aspect of social practice".[40] At worst, if these aspects were to be ignored, the risk is that knowledge will be regarded as a commodity, and the learning individual an object upon which certain knowledge is to be transferred. This can also manifest itself in conflicts between "learning to know, and learning to display knowledge for evaluation".[41] In other words, there are a number of quandaries in which the pursuit of professional knowledge is now pursued, for instance that of conflicting measures of quality, "one based on correctness, the other on practical experience".[42]

In principle, the type of knowledge referred to here is the knowledge of familiarity, which we extract from a culture, or social context. Tacit knowledge relates more broadly to the limits of systematically verbalising human knowledge. In other words, there are intangible qualities of proficiency, what Michael Polanyi has referred to as that "we can know more than we can tell".[43] In that way it also signifies a constraint within professional activities with regard to the explicit articulation of skill, a gap that management sciences are seemingly trying to bridge:

> An immense library of distinguishable situations is built up on the basis of experience. [...] We doubtless store many more typical situations in our memories than words in our vocabularies. Consequently, such situations of reference bear no names and, in fact, seem to defy complete verbal description.[44]

Typically, acquiring skills begins with the following of certain rules, and a key characteristic about rule-following is that the act of following a rule in a given situation, in a particular setting, defies complete articulation. Within an establish praxis, rule-following takes place pursuant to certain representative styles or patterns. Likewise such *intransitive* rules are embedded in action as part of practice. It is in how we follow a rule, or *act* in situations of practice, that we are able to fully express our professional knowledge.[45]

> An elementary factor in what is usually called experience is that experienced individuals are able to manage a variety of improbable or unexpected situations. Just as it is not possible to define an unexpected situation beforehand, in all likelihood our ability to manage the situation should not lead up to appreciation that we must be able to articulate the knowledge that makes this management of the unexpected possible.[46]

There is a compulsive element in the rule-following within certain praxises; it usually has its own tacit coherencies and foundations that are developed over time between people of the same work place. When someone has become experienced, he or she enters the phase of a potential surpassing of rule-guided *knowing that*; this is not an act of objection but learning to master a great variety of situations in a manner that will enable one to extend the boundaries of these rules, whenever the circumstances require us to.[47] Thanks to this *open* character of rule-following, that following a rule omits variation, and that we are able to act upon our judgements and analogical thinking, being a skilled practitioner implicates being able to manage the unexpected, "the ability to recombine fragments of past experience into novel responses".[48] This is another side to the critical thinking aspect of professional skill, the kind of interpretive skills needed to address practical situations and tasks on the bases of established knowledge as well as personal experience.

There are often different ways of implementing a certain objective. Responsibility, in that sense, cannot only relate to situations where the outcomes of our decisions and actions are clear and evident. It is bound to implicate also the management of uncertainty; to respond to shifting conditions; to accommodate change and discontinuity. If knowledge is imagined as something that we are able to "transfer" between different individuals and also between generations and organisations, this imposes an insight into the fact that experience-based knowledge becomes a *new* knowledge among those who receive it. As important aspects of human proficiency cannot be written down, such transmission of knowledge must be predominantly tacit.[49]

This also links with modern science, particularly the new interdisciplinary research field of neuroeconomics, where it is established that the utility perceived when the outcome of a person's actions, choices and decisions is actually experienced is a crucial factor in human learning and decision making. It seems as if these experienced utilities are computed into the long-term memory of the brain, more precisely the orbifrontal cortex, to guide future actions, enabling good decision-making under conditions of uncertainty:

> Therefore, it can be hypothesised that we learn about decision utilities for different actions through trial and error based on the experienced utility we obtained when taking those actions (or similar actions) in the past.[50]

In other words, "the ability to use past experience to guide future choices arises from trial and error learning, that is, by repeatedly taking particular actions and then observing the outcomes of those actions".[51] By means of analogical thinking, professionals are able to produce risk-sensitive decision making, based on previously encountered utilities and associations. It is also an indication that new employees are likely to be less able in managing the unexpected, due to a lack of "hands-on" experience. If the same system of rules and instructions is put into the hands of an expert or a beginner, the amount of knowledge will not be the same. This does not necessarily help us to sleep better at night, but it can give us important clues to why the knowledge of experienced personnel is vital in the development of safety cultures. The acquired understanding underlying the capacity of good decision making does not lie in the assimilation of theory or abstraction but in knowledge relating to sensory impression and experience.

In his writings, *Rameau's Nephew* and *The Paradox of the Actor*, Denis Diderot (1713–1784), Chief Encyclopaedist, addressed this epistemology in the process of Enlightenment. The paradox of the actor means that actors are never quite consistent in playing their roles, an analogy also applicable to skilled professionals in other areas of expertise.[52] His or her performance is deepened through the virtues of repetition, in playing their part over and over again. If everything was to be articulated by means of scientific method, Diderot anticipated a development where humans would no longer have access to their sensory experiences, and the kind of ingenuity and perception spawned thereby; a peculiar smell, a false sound, something that does not *feel* right.[53]

This also represents a counterweight to the abstract classifications of Enlightenment, the belief in the powers of rational enquiry "to understand, to order and to finally master and control the objects of human knowledge".[54] What F. W. Taylor, himself a supporter of piecemeal transformation, did not pay much attention to was indeed those tacit dimensions of knowledge that defy complete articulation. This is the kind of knowledge that is liable to erosion. He had a seemingly limited appreciation for the kind of conditions operatives need to acquire and retain skill and ability in the long run. In his view, the main incentive to perform good work was simply *money*, and he was convinced that detailed instructions, describing the optimal way of performing each task, ought to bring the best out of each operative.[55]

In the operating of nuclear power plants, predictability and enforced standardisation have become next to synonymous to issues of quality. Meanwhile, the experience-based knowledge that make up the basis of rules and instructions is liable to corrosion, as future generations might interpret these instructions to the letter.[56] Although regulatory authorities like the Swedish Radiation Safety Authority (SSM) may well desire an equal relationship to plant management, they now seem to push these organisations towards increased formalisation, with new demands for explanation, proof and evidence of action. This "narrowing of vision", making a higher degree of schematic knowledge possible, has also led to a process in which the analogical and critical thinking of human experts are in some measure deprived.[57] Besides, trying to speed up the learning of plant personnel, which has been called for at certain power plants, is likely to impoverish the quality of operating organisations in the long term.

Within many organisations, the influence of Knowledge Management-type methods and principles has superseded that of Fordism and Taylorism, which are now outdated. Although tools and techniques may vary, in both traditions a main target has been to foster a capacity for quicker and more "effective" organisational learning, promoting the objectives of transparency and codification, in creating "common knowledge"; to make personal knowledge accessible via certain information systems, experience systems or "expert systems".[58]

The idea behind expert systems is to formalise the knowledge of the expert, gathered from accumulated experience. The goal is enhanced reliability and transparency, as well as to make the practical knowledge systemised and more easily accessible to newly recruited staff.[59] In a Taylorist tradition the purpose is also to create new knowledge, transcending the old one. Thus formal work place education is assumed to be the main source of learning. Within the tradition of Knowledge Management, sharing knowledge between individuals is typically regarded as non-problematic. And although it discriminates between different *types* of knowledge, the main objective has nonetheless been to convert tacit knowledge into explicit knowledge, making it portable, exchangeable, and less dependent on specific groups or individuals: Dissemination of tacit knowledge within organisations is set to occur by means of codification or through models and metaphors, although no one is assumed to have exclusive responsibility for the creation of common knowledge, as within a Taylorist tradition:

> Underlying the theme of conversion of knowledge, from tacit to explicit, (in Knowledge Management) there is hidden the same desire observed in Taylorism, namely that of formalization and exteriorisation of personal knowledge, for the purpose of converting it into organizational knowledge. Authors often insist on the importance of this kind of knowledge, but they also emphasize the difficulty of managing it.[60]

To make knowledge less dependent on the individual makes for an attractive package to many work place organisations. Along the lines of these approaches to quality, knowledge becomes a matter of selection and will be regarded as short-lived. Still, Knowledge Management-type theories and attempts to formalise the knowledge of organisations usually go out of fashion once enough establishments come to realise that it does not work, that it becomes too expensive or ineffective. In many areas of work it has nevertheless had a significant impact. As the panel discussion at 3rd WANO Knowledge Management Workshop in Paris 2010 suggests, Knowledge Management is an area of many strategies and banners: "To have the right knowledge in the right place, for the right people at the right time" is suggested by John Day, Head of Knowledge Management at Sellafield in north England. More specifically, Knowledge Management ensures that operatives have the right knowledge *at hand*. Julio Benavides, Training Manager at one of Centrales Nucleares Almaraz facilities outside of Madrid, takes it one step further. He argues that Knowledge Management is all about "how to achieve that the guy, for instance an operator, makes the right decision in the right manner, at the right time". In other words, to make sure that plant personnel have the right instructions, the proper training or work place education, in order to make the right decisions, in the right manner and at the right time. David Gilchrist, Head of Nuclear Operations at Italian Enel, stresses that, within the Nuclear Power Industry, "basically everything becomes knowledge" that can be surveyed and disseminated, but that it is considerably more difficult than that. What can and should be done is "to systematise as far as possible" the things that can be dealt with, in order to delimit what cannot be controlled and monitored. Many participants in this workshop believe that there is a tendency within today's nuclear industry to overemphasise the merit of formal knowledge and education. Previous generations have learned more independently, whereas today's education is oriented towards enforced standardisation and the management of information.[61]

The impact of Knowledge Management-type methods and strategies points to the fact that various sectors of industry are indeed affected by trends and tendencies of society. In that sense, the Nuclear Power Industry is not as secluded as one might expect. This is exemplified and accentuated further in the NAIIC report, in which cultural impacts of general society, such as the lack of social development and critical thinking, are getting a major part of the blame for the misadventures at Fukushima. Arguably, these cultural tendencies have also contributed to the fact that the knowledge ideal of a model has been increasingly favoured, often at the expense of experience-based, manual knowledge; knowledge that evolves over time, enabling the work of "reflection and imagination".[62]

In the longer run, however, if pushed into this kind of narrowing of vision we are forced to reduce our judgements and critical thinking. In other words, we must also draw attention to the knowledge that also makes it possible to manage the unexpected. In high-risk activities, the training of operatives should have the purpose of enabling them to operate in settings that are particular and undetermined; where some facts are unknown or when normal conditions no longer prevail.

This is the kind of rule-following through which someone learns to carry out a specific task in the first place, emulating a model, or role model, gradually learning to take significant variation into account.[63] Even if skilled practitioners are able to describe a great deal of what they know, these are some of the processes that defy complete articulation. From an analyst's point of view, the objective is usually to break up relevant skills and qualifications, to divide work into separate tasks and to identify specific tasks to be trained, to extract and differentiate; to define certain tasks as qualified and others as unqualified or procedural. Professional knowledge is thus divided into building-blocks that can be detached and reassembled, as adjustments are being made to one block or another. This, essentially, is the logics of pre-packaged education, of re-engineering and technological change. In the interdisciplinary, practice-oriented research field of Skill and Technology, rather, professional knowledge is conceived as a cohesive entity that cannot be decomposed.[64]

Yet in theory our knowledge, as for instance in an occupation, can be divided into three parts: the knowledge we acquire by practising this profession (knowledge expressed in skill), the knowledge we attain from watching predecessors, as well as by exchanging experiences with colleagues and fellow workers (knowledge of familiarity), and the knowledge we learn by actually studying the subject by way of formal education (conceptual/propositional knowledge). The relationship between these three aspects of knowledge can be delineated in the following manner: "We interpret theories, methods and rules by means of the familiarity and experience we have acquired through our participation in a practice."[65] From this we can also conclude that there are intrinsic associations between different types of knowledge related to professional activities. For habits, attitudes and experiences to develop into skill, practitioners must work with resistances; befriending ambiguity, enabling the mastering of techniques with "minimum force".[66]

This, in essence, goes back to Aristotle and the virtues of good life. Accordingly, practical experience (of actions) is more important than theory in order to become skilled, or *virtuous*. On the bases of these three aspects of professional knowledge there is strong potential for the formation of considered judgements, the *synesis* of Aristotelian ethics. Unfettered judgements are bound to be predominately qualitative and subjective, though they are rooted in professional praxises and in experience.

Professional knowledge is both individual and collective, developing from theory as well as from the experiences and (collective) trial and error of individuals working together. Yet the acquisition of skill does not pursue along the lines of linear development; it rather detours and encounters various forms of resistances. Novices and apprentices must learn to recognise these resistances, and to sense the difficulties encountered instead of aggressively "conducting war against them".[67]

The world is still unpredictable and work-related problems are not always descriptive, or well-defined. Managing certain problems or situations may well involve enforcement of a rule, or detours from established praxis. Likewise professionals in safety-critical activities, and in all areas of work, must be given the opportunity of evaluating the quality of their own practice.

Notes

1 Spelplats 1.2009, p. 40; Spelplats 3.2009, pp. 65–67; Spelplats 2.2014, p. 37.
2 Perrow, C. (2011): p. 142.
3 Berglund, J. (2013): pp. 76f.
4 Spelplats 1.2009, p. 20 and Spelplats 3.2009, pp. 69–70; 79.
5 Ericson, M. and Mårtensson, L., "The Human Factor?", in Grimvall, G., Homlgren, Å, Jacobsson, P. and Thedéen, T. (2010): p. 250.
6 Berglund, J. (2013): pp. 44f.
7 Stensson, P. (2014): p. 25.
8 Cf. Skinns, L., Scott, M. and Cox, T. Eds. (2011): pp. 85f.
9 Dialogue Seminar, 13 September 2012, at the KTH Royal Institute of Technology, with participants from the Swedish Nuclear Industry.

10 Berglund, J. (2013): pp. 42f.
11 Master of Skill and Technology seminar, May 2014, Linnaeus University, with participants from the Swedish nuclear industry.
12 Dieckmann, P., Molin Friis, S., Lippert, A. and Østergaard, D. (2009): "The Art and Science of Debriefing in Simulation – Ideal and Practice", pp. 287–292.
13 Berglund, J. (2013): pp. 81f.
14 Woods, D., Dekker, S., Cook, R., Johannesen, L. and Sarter, N. (2010): p. 130.
15 Cf. Rasmussen, J. and Svedung, I. (2007): pp. 13–16.
16 Dialogue Seminar, *Safety Culture across Organisational Borders*, 4 March 2013, at KTH Royal Institute of Technology.
17 Ericson, M. and Mårtensson, L., in Grimvall et al. (2010): p. 253.
18 Stensson, P. (2014): p. 192.
19 Cf. Weick, K. and Sutcliffe, K. (2007): pp. 83–84; 150–151. Churchill quoted on p. 83.
20 These comments are summarised from a Dialogue Seminar series presented in Berglund, J. and Leijonberg, A. (Eds.) Spelplats 1.2012.
21 Woods, D., Dekker, S., Cook, R., Johannesen, L. and Sarter, N. (2010): p. 77.
22 Cf. Taleb, N. (2013): pp. 8–10; 119–129.
23 Cf. Woods, D., Dekker, S., Cook, R., Johannesen, L. and Sarter, N. (2010): pp. 76f.
24 Corradini, M. and Klein, D. (2012): p. 28.
25 Engström, D. (2015): *Att styra säkerhet med siffror* (Managing Safety by Numbers). Master Thesis in Skill and Technology, Linnaeus University. Diana works at the Department of Nuclear Safety at the Swedish Nuclear Fuel and Waste Management Company (SKB).
26 Sennett, R. (2008): pp. 230–238.
27 Ibid., p. 247.
28 Cf. Rasmussen, J. and Svedung, I. (2007): pp. 13f.
29 Taleb, N. (2013): pp. 3–7; 21–22; 57–59; 163f.
30 Ibid., p. 71.
31 Ibid., pp. 61–62.
32 Ibid., pp. 45–46; 72–73; 212–213; 334–339; 354–356.
33 Cf. Sennett, R. (2012): pp. 17–18; 127–129.
34 Taleb, N. (2010): p. 135.
35 Taleb, N. (2013): pp. 375–386.
36 Hughes, W. (1996): *Critical Thinking*, pp. 21–22.
37 Ibid., p. 161.
38 Spelplats 1.2011, pp. 40–42.
39 Taleb, N. (2010): p. 181.
40 Lave, J. and Wenger, E. (1991): *Situated learning – Legitimate Peripheral Participation*, p. 31.
41 Ibid., p. 112.
42 Sennett, R. (2008): p. 52.
43 Polanyi, M. [1966] (1983): *The Tacit Dimension*, p. 4.
44 Dreyfus, H. and Dreyfus, S. (1986): *Mind over Machine – The Power of Intuition in the Era of the Computer*, p. 32. The authors in fact intend to separate the conceptions of "skill" and "knowledge". Following this line of argument, the concept of tacit knowledge is ruled out. What cultivates skill, if following the arguments of this book, is *tacit experience* rather than tacit knowledge.
45 Johannessen, K. S., "Rule Following, Intransitive Understanding and Tacit Knowledge: An Investigation of the Wittgensteinian Concept of Practice as Regards Tacit Knowing", in Göranzon, B. Hammarén, M. and Ennals, R. Eds. (2006): pp. 268–269ff.
46 Janik, A. (1991b): p. 117. My translation.
47 Ibid., pp. 112–116. The division of knowing-that and knowing-how is attributed to British philosopher Gilbert Ryle.

48 Weick, K. and Sutcliffe, K. (2007): p. 3.
49 Cf. Polanyi, M. [1966] (1983): pp. 60–61.
50 O'Doherty, J. P., "Decisions, Risk and the Brain", in Skinns, L., Scott, M. and Cox, T, Eds. (2011): p. 47.
51 Ibid.
52 Göranzon, B. Ed. (1995): *Skill, Technology and Enlightenment: On Practical Philosophy*, p. 6.
53 Josephs, H. "Rameau's Nephew: A Dialogue for the Enlightenment", in Göranzon, B. and Florin, M., Eds. (1991): *Dialogue and Technology – Art and Knowledge*, p. 151.
54 Ibid., p. 148.
55 Taylor, F. W. (2007): pp. 31–33; 105–112.
56 Spelplats 3.2014, p. 38.
57 See Berglund, J. (2013): pp. 81–110; 201f; Crawford, M. (2010): *Shop Class as Soulcraft*, p. 44–45.
58 De Vos, A., Lobet-Maris, C., Rousseau, A. and Wallemacq, A. (2002): "Knowledge in Question: From Taylorism to Knowledge Management", p. 8, referring primarily to Nonaka, I. and Takeuchi, H., *The Knowledge Creating Company*. Oxford: Oxford University Press, 1995.
59 Josefson, I. (1995): "A Confrontation between Different Traditions of Knowledge – An Example from Working Life", in Göranzon, B. Ed. (1995): pp. 261f.
60 De Vos et al. (2002): p. 11.
61 Seminar at WANO 3rd Knowledge Management Workshop, 17 March 2010, WANO Paris Centre, Neuilly-sur-Seine.
62 Sennett, R. (2008): p. 295.
63 Sennett, R. (2004): *Respect in a World of Inequality*, p. 232.
64 Perby, M-L. (1995): pp. 188–189ff.
65 Göranzon, B. in Göranzon, Hammarén and Ennals Eds. (2006): p. 190.
66 Sennett, R. (2008): p. 215.
67 Ibid., pp. 226–230.

6   The Concept of Quality Revisited

6.1   Indirect Approaches to Quality

In many areas of work there is a need to broaden the exchanges of knowledge and experience, beyond the current uses of training and formal education. As for high-risk activities, there may also be a need to recapture the conception of quality, for instance as regards its close association to experience-based knowledge. In the pursuit of safety, the proficiency of plant personnel is evaluated against the consistency of meticulously elaborated models of the work content, which has a strong impact on the focus of training and further education. Still, replacing a knowledge ideal based on analogical thinking with one more intimately related to the concept of a model is likely to lead to some sort of reduction of critical thinking and reflection in the longer term. Safety-critical work is by no means some laissez-faire enterprise; certain restrictions to the degree of freedom that a decision presents to the decision-maker is bound to take place as operatives adjust to the support (and constraints) of carefully worked-out procedures, rules and instructions. At best, standard procedures will represent the safest and easiest way for operatives to ensure quality and to manage the expected, while managers are seeking to control the process, not the person managing it.[1] However, to perform a certain task over and over again is not equivalent to a narrowing of thinking. Something evolves over time in daily business activities, while doing the same or similar types of work; a refined capacity to detect problems, solve problems and to accommodate ambiguity and uncertainty is nurtured in the process of slow learning, the work of reflection and analogical thinking.

In traditional behaviouristic methodology, on the contrary, learning equates to positive changes in observable behaviour. As philosopher Hans Skjervheim has pointed out, when approaching human activities in this manner, actions, gestures, claims or propositions are regarded as *facts*, as opposed to actions, gestures, claims or propositions, which can be taken up for discussion, engaging us in dialogue. In various studies of professional skill, learning is rather considered a result of reflection upon concrete experiences, personal and joint. In the progress of this research field it has been assumed to be an epistemological mistake to explore adults' proficiency in a manner of straightforward questionnaire. Nor has a positivistic approach of external observations of work been deemed feasible; stereotypes and mock-ups often stand in the way.[2] Categorising other people's behaviour is only ostensibly simple:

> Even for highly constrained task situations such as nuclear power opera-tions, modifications of instructions is repeatedly found, and operators' violations of rules appear to be quite rational, given the actual workload and time constraints. One implication in the present context is that fol-lowing an accident it will be easy to find someone involved in the dynamic flow of events who has violated a formal rule just by following estab-lished practice. He or she is therefore likely to be exposed to punishment. Consequently, accidents are typically judged to be caused by "human error" on the part of the train driver, a pilot, or a process operator.[3]

There is also the dilemma of the *hindsight bias* in judgements of what caused the accident or event in question. After an accident has occurred, we know the outcome and such information tends to bias people's opinions about the processes, actions or non-actions which led up to the unfortunate outcome. In hindsight, our reactions, or the reactions of analysts and investigators, are typically as if this knowledge of outcome was also observable to operators at the sharp end of the system, wondering why they did not anticipate it. This oversimplifies the situation facing the practitioners, and will obscure many latent factors and processes affecting their genuine response, making the correct choices of action seem "crystal clear". Still, design and application are two different things once reality goes its own way. Ironically, these are the kinds of reactions and analyses that force utility companies to establish more detailed and comprehensive rule-systems and procedures, and to ensure that operators follow these procedures rigidly. For that reason, our reactions to accidents are not always prone to regard failure simply as opportunities for learning. It rather tends to solidify the inherited caste system of knowledge and credibility, and the conception of the human factor as the main source of error within technological systems, forcing managers not to get their hands dirty but to impose further constraints, and system engineers to use more and more technology to reduce the number of action alternatives available to the practitioners who are operating the process.[4]

In other words, this idea of a "human error problem" within socio-technological systems every so often leads to increased automation and for-malisation of work. But automation and the introduction of new technology in the work place also makes systems more complex, affecting "the potential for different kind of erroneous actions and assessments". It can reduce the risk of some kinds of failures, while at the same time creating or increasing the potential of others:

> Decisions and actions followed by a negative outcome will be judged more harshly than if the same decisions had resulted in a neutral or

positive outcome. Given knowledge of outcome, reviewers will tend to simplify the problem-solving situation that was actually faced by the practitioners. The dilemmas, the uncertainties, the trade-offs, the attentional demands, and double binds faced by practitioners may be missed, or under-emphasised, when an incident is viewed in hindsight.[5]

An instruction, or task description, is often unreliable as a model for judging behaviour during actual work. But there are still ways for organisations to advance from the dominant role of enforced standardisation. Going forward, I will discuss some alternative approaches to quality work and further education, which can empower people with insight into their own experience as well as others; to foster analogical and critical thinking, with a key role for dialogue. Besides, this can also work as a counterweight to the separation of "knowing" and doing.

Experience is said to be transferred, or matured, into knowledge when we reflect upon it, and reflection surely has an impact on experience. To Diderot the unmet need for reflection was a sort of *illness*, which had to run its course. Yet the implication attached to the conception of reflection can vary considerably between different contexts of learning. There are several, partly overlapping, images of reflection, each with significance to professional practices: (1) The *dedoublemént*, or "role splitting" of contemplating something from dissimilar positions; (2) analogical thinking between examples, situations or experiences that are non-identical; (3) the act of "bending one's thoughts back"; (4) the experimental "reflection-in-action" of Donald Schöns reflective practitioner; (5) the "puzzle-solving" process of fitting pieces together, or looking for different solutions to the same problem; and (5) the dialogical movement of "criss-crossing a landscape", as opposed to walking down the same trail of thought, or topic, where digressions are taken seriously.[6]

Not all thinking is reflective, and some of the thinking that goes into learning is rather unreflective and intuitive. According to philosopher Allan Janik, reflection is neither entirely conscious nor unconscious. It is a sort of hermeneutic activity, a way of "finding orientation" in situations of ambiguity, adversity or disorder; where our current practices no longer seem to function; where we no longer understand our situation.[7] To reflect we sometimes need to go outside of ourselves, by means of a narrative, like a piece of literature, which we can use as a mask or mirror. Reading and writing can structure out thoughts and foster the reflective processes of learning, but can also illuminate something of the complex character of human experience.[8]

Hence, experience is not the only criterion of skill; to widen our perception and thinking capabilities are equally important. But reflection does not always comes naturally; production pressures, re-engineering of work and other forms of insecurity can inhibit us from creating a fragmentation of time and generating a personality type "constantly in recovery", incapable of constructively challenging their everyday practice.[9]

Many times, our imagination can supersede what we lack in experience. But imagination will often require some sort of impulse; a work of preparation, such as the incitement of literature. As C.P. Snow has pointed out, as we read "our imaginations stretch wider than our beliefs", unlike if we construct "mental boxes to shut out what won't fit".[10] Moreover, in literature and in books there are portrayals and sequences of human experience and memories that are both particular and universal in the context of analogical thinking. Whatever impression or image we find appropriate, the process of reflection does not always come naturally. Rather, it often needs to be invigorated by means of reading and writing and in the course of dialogue; possibly even through a change of environment.

Humanistic knowledge on the basis of the humanities, along with the social sciences, are essential to individual and collective self-understanding, but they are not always given a high profile among policy makers. This is partly due to the high profile of science and technology in modern society but also a result of their "self-encapsulation" – their struggle in finding a strong relationship to society:

> Our assumption all along has been that there is not anything like a fixed specific context in which humanistic knowledge is required by society but that at certain crucial points in the development of individuals, practices and institutions distressing, because wholly unforeseen problems, arise about our activities and our very identities, which can only be illuminated on the basis of human wisdom. The fact that the "demand" for humanistic knowledge is linked to crises complicates the problem of its institutionalization but it by no means prevents it.[11]

Humanistic knowledge can help us make sense of the particular, of experience embedded in circumstances that can shift in unexpected ways. It is characterised by personal insight as well as collective knowledge entrenched in human practice; it is gained from experience of doing things and articulated through reflection upon experience: "In order to *capture* tacit knowing you need to reflect systematically upon individual and group experience and to articulate that reflection."[12] Karl Weick and Kathleen Sutcliffe propose that high-risk organisations should make use of such vicarious experiences like, for instance, Churchill's protocol after the invasion of Singapore, to use it as a sort of impulse to discuss disruptive events in their own practice and as an evocative example or analogy, but also proactively so as to foster risk awareness and critical thinking.[13]

Analogical thinking is the essence of experience development. To energise such reflection there usually has to be some sort of recognised problem with the routine, or other aspects of practice; an accident or complications regarding the implementation of new technology. The suggestion put forward in the NIRA evaluation of the 1980s, to counteract any untoward consequences of technological change proactively within the education system, necessitate this kind of reflexivity. Sometimes, we need to reflect more deeply upon the conditions that make "excellent practice" achievable, to dig deeper into the context in which knowledge is utilised in practice. How, then, is it possible to reflect yourself through failure?

As human experience is by and large *visceral*, humanists can provide us with vicarious experience of, for instance, typical human problems of conflict or failure, which can give us important insight without us having to actually live through these experiences ourselves. But the trick is also "to make that knowledge available to the person(s) in distress" at just the right time and in the right manner, to stimulate the capacity for analogical thinking. If we consider all knowledge formal or explicit, we are likely to be less equipped in dealing with uncertainty as soon as it arises concretely. In a tradition of liberal education, the humanities is as much about grammar and verbal expression as the evaluation of data and pattern recognition, all of which have teachable aspects. As academic disciplines they have been "the basis upon which higher education rests", in supplying scholars with the tools of critical thinking that are aimed at providing "a general background" for pursuing higher studies. Arguably, a lot of the future potential of humanistic knowledge and its application lies in continuing further education; to provide organisations and society as a whole with professional re-education, or re-orientation, with the support for reflection upon the complexity of human experience. For instance, a gifted engineer that has been promoted to manager will not need so much more of the technological knowledge that got him or her promoted in the first place. Instead, this person is most likely dependent on another type of competence and preparation in coping with issues of leadership and trying circumstances; problems related to professional knowledge and experience that can be dealt with "reflectively in dialogue".[14] This is the kind of education that partly depends on the experience you bring in.

In other words, one way of addressing this is by means of training and further education. Not as a substitute for the knowledge we acquire from (other) education, or from practice, but as support, reinforcing the tacit dimensions of knowledge rather than undermining them. With the Dialogue Seminar Method, reflection is brought about outside the work situation. In the reflective activities promoted through this methodology, discussion on skill and quality will take its departure from examples and analogies, as opposed to general models and abstractions. Reflecting in the process of dialogue, learning is also seen as arising from encounters with differences. For professionals to effectively make sense of their everyday practice, challenges and experiences, to nurture reflection and critical thinking is deemed more important and achievable than the knowledge conversion promoted in management sciences like Knowledge Management and Taylorism.

The endeavour is to trigger those involved to reflect more efficiently upon the things we know *until we are asked about it*; to extend rather than to restrict the imaginative thinking of skilled professionals. Rather than zooming in on a single problem, experience development allows for digressions; "zooming out", opening up for diverse opinions and experiences to get through, which may then converge into fruitful criticism.[15] Potentially, this is also a source of critical thinking. As for any organisation, if there are no detours or digressions from standard procedures, the development of practice is at some point bound to deteriorate. Contrary to the principles of Taylorism, it is often the "propensity to variation" that is the true principle of progress.[16]

To address issues of quality in an evocative way we are dependent on shared ideas and assumptions, in order to see and identify the problems and difficulties in the first place. In the case of SAAB Combitech, experience development has become a method for reflection around new projects, such as the development of new products and services. Promoting lateral thinking, this is also a means for highly competitive engineers to become more collaborative, as well as to come to terms with and appreciate the notion of average in engineering. It is a Learning Lab, so to speak, utilised in the continual education of newly recruited staff, but also among experienced engineers, "gathering experiences from a completed project, establishing a common language in a newly formed group".[17]

In this case, different experiences are developed, not transferred, between people of the same work place, or in mixed groups of people doing different kinds of work, sharing a narrative or example, to make way for dialogue and analogical thinking. Instead of mainly adding new rules and routines top-down, the alternative may be to nurture a mistake, incident or let-down within a certain praxis, section or department of an organisation, paving the way for reflection on encountering with situations where one's knowledge, or praxis, have been tried and challenged. Experience development is a means of making this reflection and learning more conscious; to make experience accessible for reflection; to capture reality by means of examples rather than models:

> It also involves creating a meeting-place in the organisation, in which the experiences of others have the opportunity to expand one's own horizons and make them more complex. It should be recognised that the term meeting-place has a twofold meaning: A physical meeting-place, certainly, a location where conversations can take place. But it is more a question of creating the condition in which experience can meet experience: bringing to life the vital examples that set tacit knowledge in motion. We should also be aware of the twofold meaning of the term experience transfer (or development). It means learning from the experience of others, but also qualifying one's own experience. In both cases, reflection has an impact on experience.[18]

In other words, the idea behind this methodology is to create a more reflective practice. To high-risk organisations this can be a means of providing novel perspectives, to formulate new thoughts and to enhance the overall risk awareness of operatives and other members of staff. In comparison the idea of so-called Quality Leadership Practice, or Quality as Empowerment, is to organise people in small-group improvement activities such as Quality Circles, which can be applied to identify and unravel problems in their own work, presenting bottom-up solutions to management. To the Japanese, this is part of *kaizen*, which literally means small steps of good change, or continuous improvement. The concept of *kaizen* also has a double meaning: Firstly, it is closely associated with the manufacturing industry, and the strict process improvement and waste elimination of companies like Toyota; to reduce costs and make the overall production process slimmer, or *lean*, aiming at "eliminating non-value-adding activities of all kinds".[19] Secondly, its original meaning is related to the essence of working together for the benefit of learning, communication and skill development. Problem solving is often a convenient activity for newly formed Quality Circles, while *kaizen* activities also can help managers keep in touch with what practitioners actually do. As quality researcher David Hutchins puts it, "everyone likes quality", but we may attach different meanings to the same concept. Enabling its associates to congregate around some of the more imperative problems in their own areas of work, such as inadequate job instructions, housekeeping problems or Risk Identification, the principle of such small-group improvement-type activities is also to influence practice, prompting a gradual transformation in job design:

> To involve people successfully it is necessary to modify the Division of Labour method to be able to incorporate the principles of *craftsmanship*. In other words, it is necessary to introduce the concept of "self-control" to reintroduce craftsmanship to groups or teams of people.[20]

Self-discipline is as important as autonomy to the principles of good craftsmanship. The advantages of the Division of Labour approach can be combined with craftsmanship by training small groups of personnel, who do the same or similar work, in the identification of weaknesses in their part of the organisation and in problem solving, with the purpose of nourishing practical proficiency in order to facilitate managers in making a linkage between corporate targets and concrete situations of operation. The overall result can only be improved if workers are delegated responsibility to perform good work and to nurture mistakes within their praxis instead of setting up some sort of external inquiry, which may not lead to the desired improvements within the organisation.[21] Besides, leanness does not always work well with safety-critical activities as it tends to strip the organisation of resilience when managers eradicate "redundant" functions and positions along with the experience and proficiency that goes with them.[22] Creating a reflective practice, or Quality as Empowerment, is also a way of motivating individuals that have reached a higher level of skill; to make better use of their experience, and to counteract the separation of "knowing" and doing:

> Different people will form different judgements in the same situation, not just because they have different objectives but because they observe different options, select different information and assess that information differently: and even with hindsight it will often not be possible to say who was right and who was wrong. [...] the skill of problem solving frequently lies in the interpretation and reinterpretation of high-level objectives.[23]

In other words, the realisation of high-level objectives, such as safety, intermediate goals and plain actions, are best approached indirectly. There is no science of decision making, which could be managed rationally, leading every conscientious person to the same answer. We do not become better decision-makers necessarily through a higher level of precision and specification a priori of the challenges we encounter.[24] Decentralisation of the clarification and solution processes of complex problems also requires empowerment of local knowledge. Moreover, these indirect approaches to quality discussed here (Quality Circles and the Dialogue Seminar Method) may well encourage the kind of reflexivity promoted by Michael Power and others (see above), and a sensitivity towards what make certain practices legitimate, not necessarily because they have been successful in achieving goals or receiving good rankings by external reviewers. Initiatives like these can also counter-balance the risk that formal education does not cover the actual needs of today when it comes to learning. Extensive formalisation generates new risks, *other* risks, in terms of long-term consequences; not what will happen tomorrow. To high-risk organisations it is a challenge not to end up single-mindedly promoting error prevention, at the expenses of payoffs in terms of learning and knowledge development:

> If revision only occurs when evidence is overwhelming, there is a grave risk of an organisation acting too riskily and finding out only from near-misses, serious incidents, or even actual harm. Instead, the practice of revising assessments of risk needs to be an ongoing process. In this process of continuing re-evaluation, the working assumption is that risks are changing or evidence of risks has been missed.[25]

Critical thinking at all levels of the organisation is vital to the build-up and maintenance of a dynamic safety culture. Besides, there must be channels for turning such critical thinking into a questioning attitude and vice versa, the kind of cultural factors or "collaborative interchanges that generate fresh points of view or that produce challenges to basic assumptions".[26]

Interchanges like these, for instance between managers and operatives, can be developed to identify areas of work in which human capacities can be more fully engaged. Still, Japanese companies are often very standard-oriented, and Quality control also means variability control. Typically, small-group improvement activities are a way of balancing the Taylorist tendencies of Japanese management practices; to involve more members of staff in the process of thinking and problem solving. In that way, their standards are continuously reviewed and upgraded. Yet standard procedures are commonly regarded as "the most efficient, safe and cost-effective way of doing the job".[27]

As Japanese companies became market-leaders within many sectors of industry, there has been growing interest in their strategies, ideas and principles of management. But few western approaches to Lean Manufacturing and Total Quality Management (TQM) seem to have fully recognised the significance of autonomous, small-group improvement-type activities, or Self-Managed Workgroups, essential to the competitive successes of many Japanese corporations.[28] In that way, many western companies and consultancy agencies have rather made selective use of Japanese experience.

6.2 Seeing Different Worlds

Japanese companies employ millions of people abroad and there have been many attempts to transfer Japanese *kaizen* activities to overseas plants in China and other countries. But, on many occasions, there have been difficulties in such activities taking root in organisations outside Japan. What then is it in the nature of these activities that are so difficult to transfer? A driving force in small-group improvement-type activities is fostering knowledge development within groups of people with different experiences, and to reduce the social distance between different categories of employees. Still, the meaning of praxis is situated in local contexts. There is also a social context in which practitioners cultivate their own language, styles and patterns of action and cooperation. The exchange of *kaizen* ideas can thus be cultivated through creating forums within the organisation, attended by employees on a voluntary basis. Lacking a tradition of democratic dialogue, in countries like China such endeavours have not necessarily involved activities leading to the active participation and empowerment of workers. In other words, there is the conception, or the ideal and then there is application. Yet in some cases industrial plants in China have even managed to outperform main Japanese ones, not by naive emulation but by applying their own cycle of small-group improvement activities. Also, *kaizen* activities have not always been a success within Japanese companies, with some managers tending to use these forums of cooperation, for instance, to discipline workers, narrowing the scope of practice rather than nurturing learning and the expansion of knowledge based on a balanced response to a problem at the actual site (*gemba*):

> [...] a meaning is always situated in the local context and is difficult to be applied beyond the local situation. Therefore, communication that is not based on a concrete situation is likely to lead to misunderstanding.

However, this practice allows management and workers to use the concrete situation as a context for learning, and thus to share understanding.[29]

From the point of view of the supervisory agencies, an ideal situation would be to strive for an *equal* relationship relative to their counterparts in the Nuclear Power Industry, on the basis of dialogue. Arguably, there is as much risk involved for organisations who uncritically comply as for those which discard any sort of external criticism.[30] In Sweden, there is little prescribed when it comes to safety culture from the supervisory authority, except that the organisation must create fruitful conditions regarding the continuation of safety and reliability. In the case of Japan and the proceedings leading up to accident at Fukushima Daiichi, some reports also claims that the Japanese Nuclear and Industrial Safety Agency (NISA) were in cahoots with TEPCO, drawing a veil over the need for certain reinforcements. Although both parties were aware of the risk of a station blackout due to a tsunami larger than predicted in the estimations of the Japan Society of Civil Engineers, which would render seawater pumps inoperable, nothing was done because "the probability was small and other measures were in place".[31] In other words, no one had witnessed a tsunami that big in their own lifetime – they were only heard of in legends.

Rather than demanding the implementation of critical reinforcements, the NAIIC report blames NISA for stating that actions should be taken autonomously by plant personnel, and TEPCO for not completing such critical reinforcement. As was revealed, NISA also asked plant operators to write a report on this issue to explain why the consideration of a possible station blackout was still unnecessary. Regulators also had a "negative attitude toward the importation of new advances in knowledge and technology from overseas".[32] This negativity seems to be unlike the usual behaviour in Japan, however it is assumed to reflect on the insularity of Japanese culture. Similar criticism is also put forward by INPO,[33] claiming that, from the 1980s onwards, Japanese utilities and vendors decided to depart from joint technology, as well as accident management strategies developed by the US Boiling Water Reactor Owner's group. On the other hand, in various instances "the practices and level of preparation for a severe accident" at Fukushima prior to March 2011 were similar to those found at other nuclear stations around the world.[34] Some reports have also been critical of TEPCO for operating reactors designed in the US where damage by tsunamis is unheard of, and in that sense had not sufficiently considered risks inherent to Japan.[35]

In any case, there are strong reasons to believe that the Japanese experience has significant implications for other countries as well, and to operating organisations around the world. It is not satisfactory, I think, to simply blame Japanese culture and society. Despite its insularity, for the purposes of international comparison Japanese nuclear facilities are up there with the best. Still, similar to several nuclear facilities and companies like Vattenfall, TEPCO relied strongly on contractors to fulfil a number of routine tasks, from plant maintenance activities to the supply of diesel fuel deliveries to the station. To carry out certain required actions during these extreme events, TEPCO had to request support from a number of external agencies. Even though there were agreements for such support in cases of emergency, this would ultimately add to the distraction of having to make additional arrangements in the course of the accident response. As enough contract workers were not available or indisposed to provide support under these conditions, TEPCO also had to train station workers to fulfil certain critical accident response functions and restoration tasks, for instance "how to use fire engines to support injection into the reactor and on how to operate some of the mobile generators used to restore electricity", but also to operate tankers to provide a fuel supply for emergency generators. Contractors were relied on for many restoration tasks since station workers were not trained to perform such activities. As articulated in the INPO report, in the course of these events, few contractors seemed to possess the required skills that were urgently needed. They did not have the proper training or the locally embedded knowledge vital to manage this type of high-stress, long-duration event. When I talked to people at Fukushima, reliance on contractors was apparently one of the things that TEPCO had decided to change. Moreover, there was an overload of information during the events and a lack of effective communication, making it difficult for headquarters' personnel to grasp what was happening on site.[36]

Even though compliance with rules and regulations is rational, in critical situations plant personnel also have to be able to respond to events that go beyond the frameworks of standard procedures. Moreover, in this case the relationship between regulatory agencies and industry was not so much equal as self-defeating. According to the NAIIC report, this was a partnership that in the long run mistook the promotion of safety with "the promotion of nuclear energy". Ultimately, no regulations were created, and no protective step against the risk of core damage from a tsunami was taken:

> From TEPCO's perspective, new regulations would have interfered with plant operations and weakened their stance in potential lawsuits. That was enough motivation for TEPCO to aggressively oppose new safety regulations and draw out negotiations with regulators via the Federation of Electric Power Companies (FEPC). The regulators should have taken a strong position on behalf of the public, but failed to do so. As they had firmly committed themselves to the idea that nuclear power plants were safe, they were reluctant to actively create new regulations. Further exacerbating the problem was the fact that NISA was created as part of the Ministry of Economy, Trade & Industry (METI), an organization that has been actively promoting nuclear power.[37]

In other words, national regulatory authorities in Japan were previously not autonomous and neutral to commercial and political interests. In this case, regulatory efforts were not sufficient in order to direct management priorities properly. The Commission also claims that TEPCO was looking to steer clear of responsibility in blaming the accident on the highly improbable, the tsunami, and not on the more predictable earthquake.[38] Seemingly, they complied with the demands and regulations of NISA but ultimately failed to set their own standards of safety and quality. The recommendations put forward in the NAIIC report, that plant personnel at the site ought to meet global standards in order to create a more knowledgeable organisation, are not dissimilar to the conclusions made by collaborative organisations like WANO and INPO. Given the strong critique of Japanese culture throughout the report, the request for global standards comes as a natural trail of argument. Likewise, as in the aftermath of several major accidents, the instantaneous response is every so often an aspiration for more regulation and enforced standardisation. Moreover, there are concerns regarding the methodology used to evaluate the height of a possible tsunami; there are also criticisms against the sincerity of risk assessment and critical thinking within the Japanese nuclear industry in general:

> The reason why TEPCO overlooked the significant risk of a tsunami lies within its risk management mindset – in which the interpretation of issues was often stretched to suit its own agenda. In a sound risk management structure, the management considers and implements countermeasures for risk events that have an undeniable probability, even if details have yet to be scientifically confirmed. Rather than considering the known facts and quickly implementing countermeasures, TEPCO resorted to delaying tactics, such as presenting alternative scientific studies and lobbying.[39]

Moreover, an overall culture of complacency and lack of critical thinking led regulators as well as operators to prioritise the interests of their own organisations, promoting nuclear power over public safety, creating a mutual incentive to prevent any sort of open criticism. The relationship between various participants, ultimately, did not live up to the principles of a "safety culture". Besides, the high-level objectives of TEPCO did not involve the risk of damage to public health and welfare. As the Japanese nuclear industry became less profitable over the years, TEPCO's management began to put emphasis on cost cutting, as well as increasing the nation's reliance on nuclear power. While admitting to a policy of "safety first", in reality safety suffered "at the expense of other management priorities".[40]

In view of the strong meta-criticism of Japanese society presented in the NAIIC report, or rather in the English preface, with huge public debts and notorious deflation, Japan, like many other countries, is at a point where it needs to change. With an ageing population, Japan arguably needs to increase immigration if it is to be able to revitalise its economy.[41] The implementation of Information Society in the 1970s and 1980s was essentially too naive and too radical, with an overemphasis on the virtues of new technology, and an overreliance on central planning and rational design. More importantly, it did not properly consider the distinction between knowledge and information. The Plan for Information Society, quite simply, came too close to the prospects of re-engineering, and making a decisive break with the past.

Assumptions made about entire nations and societies are bound to be marked by prejudice. As stated by Chairman Kurokawa, compared to many other cultures the Japanese have a strong inclination towards "harmony". So, when he questions the reflexive obedience and the reluctance to question authority, the commitment to sticking with the programme, the groupism and insularity of Japanese society, characterising the events of 2011 "very painfully" as an accident "Made in Japan", which could have been mitigated by a more effective response, how can this be understood?

What he means, I believe, is that the probability of flooding by a tsunami was not zero. A prominent culture of engineering may well have generated an overreliance on technology and scientific forecasting when deciding on the proper location and construction of these utilities. Furthermore, no one blew the whistle, and if somebody did no one cared to listen. There is a strong feature in Japanese culture of imitation and rule-following in a strict, regulative sense. Both the educational system and corporate cultures promote working together in groups, doing what is best for everyone. People are basically taught to think and act like that from a young age, and through the steps in life Japanese people are, by and large, pushed into compliance, which also causes many problems. When it comes to safety, perfectionism regarding technicalities is not enough. Assuming that we can know more than we can tell, there must also be a capability to intervene in order to manage the unexpected. A proactive mindset, a preparedness to address ambiguity and to respond to the unforeseen is required:

> Those responsible must make a continuous effort to raise existing safety standards. The construction and operation companies should not presume the quality of the standards of the regulatory agencies, and should not have a passive mind-set toward security and safety issues. The nuclear plant operators have the most clearly defined responsibility to prevent accidents and stop any escalation in consequential damages. In an emergency situation, the operator is required to make decisions [...]. For this reason, the operators must be competent to do so.[42]

In an imperfectly understood world, high-level aims and objectives such as safety, quality, success or profitability are necessarily complex and imprecise, and are often best approached indirectly. For many decades, small-group improvement activities have been utilised by competitive manufacturers to find the "loose brick" in other companies, which can give them an advantage against their competitors. In the Nuclear Power Industry, on the other hand, the failure of one facility is a loss of credibility to other power plants as well as to the entire industry. Rather than focusing on waste elimination and process improvement, these activities can be utilised to encourage employees to challenge the status quo of the organisation, raising matters of significance or to create a more cooperative work culture.[43]

In the case of Fukushima Daiichi, even if there were continuous improvements, there was also a "sticking-to-the-programme" mentality, a lack of critical thinking and a deficient readiness to act in situations that are undetermined. The impact of a high-stress, large-scale event like this on personal morale and decision-making capabilities should not be underestimated. Still, there are various factors that influence the unplanned interventions and social interactions of skilled professionals.

Safety analyst Sidney Dekker has launched the concept of Just Culture to delineate the kind of qualities, such as mindfulness and knowledge development, which ought to be promoted in safety-critical organisations. Human actions interacting with high-technological systems do not intend to cause damage. In contrast, highly trained and skilled professionals often come to work to do their job as well as possible. In keeping with Dekker, a Just Culture accepts the value of multiple viewpoints, and uses them to promote learning, safety and accountability. Besides, in safety-critical organisations in particular, we need to accept that there is a gap between written guidance and actual practice. Likewise, the causality of error must not be oversimplified; the complexity of real-time events are bound to be somehow contradictory. Just Cultures would also pay closer attention to the "view from below" in their own long-term interest. Potentially, there is a key role in this for organisations to facilitate a broader support for learning and knowledge development. Regarding human error as a *symptom* rather than a cause of incidents, it is in the interest of the well-being of organisations to create an environment in which practitioners can "meaningfully contribute to the context in which they work".[44] In a Just Culture, employees are treated as indispensable, not as the main source of problems:

> If professionals consider one thing "unjust", it is often this: Split-second operational decisions that get evaluated, turned over, examined, picked apart, and analysed for months – by people who were not there when the decision was taken, and whose daily work does not even involve such decisions.[45]

Also, within many work place organisations there is a tendency to mistake communication for information-sharing; where the latter is based on definition and classification, communication in a sense of dialogue is essentially as much about "what is left unsaid as said". As in the case of the establishment of Information Society, or for that matter in *our* information society, in a technological culture every so often this "open" form of communication is impoverished, for instance due to overestimation of information technologies. In civil society as well as in the professional associations of working life, this is the kind of cooperation that is essentially fading away. Most people have a capacity to handle complexity, to sense and distinguish what others mean but do not explicitly *say*, or are able to fully articulate. This is a key aspect of dialogue and a prerequisite for interpersonal cooperation.[46] Yet within certain cultures or organisations, people can be withdrawn from one another, as a result of a lack of a common language or understanding, to the point where they are pretty much using the same words but seeing "different worlds".

After the accident at Fukushima, the concept of safety culture, and *safety climate*, has been getting more attention and it has been assumed that there was a safety culture in effect at the power plant, that is at the actual site, but that it had not been built into the organisation as a whole. Maybe not surprisingly, each person's definition of safety depended on "what role or position they were in". Researchers in Japan have analysed what kind of language was used at the time of the accident and in what way discourse acts and dialogue took place, looking at, for instance, the video conferences of TEPCO. Focusing on Fukushima 1 (1F), the research group noted several irregularities regarding communication and cooperation during these accidental courses of events. Most notably, the essence of many central terms and topics tended to vary considerably between the people on site, TEPCO Head Office and the 1F off-site centre. For instance, the Japanese concept of *Onegai* can have any of the following English meanings: "Thank you in advance"; "Thank you for your help"; "I wish ... / please"; "I hope you will take good care of this"; "Please take care of the rest"; "You are urged to ... "; or "My request is ... ". Hence, the diversity of meanings attached to the concept of *Onegai* is multiplied by factors such as location and speaker, and is used to *request* something as well as to *instruct* another party. At Fukushima 1, this expression was used by plant management in a sense of requesting support from Head Office, but also in a sense of "top priority", to instruct workers on site to carry out certain countermeasures to prevent the situation from becoming much worse, for instance to ensure that the fuel of the reactor would not become exposed to danger, considering what had happened nearby at Fukushima 3. In other words, when plant management used the term *Onegai* it had very strong implications, in a sense of emergency, when addressing station workers as well as Head Office. From the Head Office, in contrast, it was used in a sense of encouragement to take action. But as it happened, as the result of a line of miscommunications, they were not able to provide enough support to arrange for deliveries of more diesel fuel to the station. As conditions were far from normal, plant personnel on site were unable to accomplish many of the tasks and measures requested from Head Office executives. On several occasions the people at the Head Office made a number of requests that they considered reasonable, but were in fact unreasonable, or even misguided, under the circumstances. Essentially, the priorities of various tasks and countermeasures were mixed up.[47]

According to plant management on site, the directives of Head Office and the measures insisted upon were too much to ask from a small workforce dealing with conditions that were anything but normal. Further, on-site decision-makers now had to take radiation levels and the concern for human life into account. Sticking to the existing structure and chain of command no longer seemed to be a convincing strategy. To prevent further damage, plant personnel had to come up with quite creative measures to respond to shifting contexts. As a result of the lack of dependable probability estimations, the impact of a tsunami so high was by and large beyond anticipation. Also, attempts to control the situation were aggravated by the different conceptions of reality, different sectors of organisation in between:

> Looking at the situation at the time the accident was in progress, analysing the use of language from multiple approaches, made it clear that aspects of the use of language differed considerably by location and function. [...] It would be difficult to say that a single reality was shared at an organisation-wide level within Tokyo Electric Power (Company).[48]

One way of coping with such perplexities would be to opt for a higher degree of clarity and precision of communication in situations of crisis and uncertainty. As was revealed in the NAIIC report, there was also a "sticking-to-the-programme" culture, causing interruption of urgent emergency responses. Typical of the corporate culture of TEPCO is a tendency to keep matters unsaid and secret when it comes to communication and information-sharing. Apparently, this has partly to do with the pressure of media, which is applicable to the entire industry. But even early in the crisis, the Japanese Cabinet Office was in contact with TEPCO to get an update on the problems it faced and the status of reactors, but information was scarce and slow in coming. The message they got was that all the reactors had survived the tsunami without any radioactive releases. Following such mix-ups of cooperation, the ongoing crisis communication between TEPCO and the cabinet was marked by suspicion.[49] Apparently, rather than making concrete decisions and communicating them to the government, "TEPCO insinuated what it thought the government wanted and therefore failed to convey the reality on the ground".[50]

The Japanese demand nothing less than complete safety, devoid of ambiguity, fostering a projection of "risk zero". This is the type of safety culture problem that occurs all over Japan. To make matters more complicated, before the accident, regulatory authorities were, to a large extent, controlled by the power companies, perpetuating a mutual objective of secrecy. In the case of TEPCO, the lack of critical thinking seems to have gone high up in the caste system of credibility, widespread among executives and risk analysts. While there has been strong focus on preventing terrorist attacks following 9/11, there has been complacency regarding natural disaster and the pursuit of continuous improvements.[51]

The directives from Head Office manifested a conception of reality in which it is possible to respond through rational procedure, while plant managers on site were faced with a reality full of uncertainty, with danger to human life, and in which support from the Head Office and external agencies were essential in preventing further damage. The lack of shared assumptions along with deficiencies of communication, meanwhile, obstructed swift and appropriate decision making.[52]

In good Japanese management practices, or in any good management practice, rather than relying on statistics and secondary information, managers will try to seek out what is happening at the sharp end of the organisation. To keep up with what practitioners actually do, allowing more open forms of cooperation in identifying complexity, root causes, countermeasures and areas of improvement, bridging the gap between top-level management and the workforce:

> Most managers prefer their desk as their workplace and wish to distance themselves from the events taking place at the *gemba*. Most managers come in contact with reality only through their daily, weekly, or even monthly reports and meetings.[53]

Often, such gaps from long-term cultural changes of work will generate a condition within the organisation which can make companies vulnerable to both internal challenges and external competition. Apparently, this was the case with TEPCO and the accident at Fukushima Daiichi. Within high-risk activities, the objectives of small-group improvement-type activities must be adjusted to the needs of specific areas of work. In other words, it does not have to be identical to the manufacturing industry.

At Fukushima, Quality Circle-type activities were part of the operating organisation on site. One of the problems was the lack of commitment and shared understanding from top management. According to the managers I interviewed at Fukushima Daiichi, in the case of TEPCO, Quality Circle functions had not had "outstanding effects" and, even if improvement activities had been promoted, they have not been regarded as a priority. There was also a lack of understanding as top-level management was apparently out of touch with core activities. Ultimately, complacency and a lack of "safety consciousness", as well as the prejudice of the nuclear department of TEPCO who thought nuclear safety had already been established and thus regarded the safety of operation as a priority of management issues, resulted in the lack of preparation for accidents. Arguably, these are as much the effects of a growing distance between decision-makers at various levels of the organisation as a problem of Japanese national culture, with top-level managers well-distanced from the daily activities carried out by plant personnel. The manufacturing industry of Japan has usually placed more emphasis on promoting bottom-up activities; within the Nuclear Power Industry there are many regulations and there are limited degrees of freedom within which the potential for change is manageable. Seemingly, the cooperative activities of Quality Circles are reduced to reporting to management on deviations, upsets and incidents, removing the potential for learning and knowledge development.[54]

In an ideal case, these activities are utilised as more open forms, or forums, of cooperation and communication; open to "negative" information and critical thinking, along with exchanges on issues of safety and quality. Although many high-profile accidents in the 1970s and 1980s were in part caused by a disorganised style of management, or lack of clarity and transparency, future accidents may well develop from other origins, such as the overprotection of systems through automation and extensive formalisation. Degeneration is usually manifested in times of crisis, for instance if top management has lost touch with core activities or made other priorities in terms of safety. Nevertheless, and unsurprisingly, the Japanese nuclear industry is now trying to unravel the problems detected during evaluations of the Fukushima Daiichi accident by means of new technology and increased standardisation, such as addressing the communication problems during the accident with benchmarking, in this case the introduction of a standardised emergency response system (ICS), clarifying the decision making and chain of command in the case of accident.

Quality Circles have high status in countries like Japan, though in a Japanese context they have arguably not been as successful at nuclear power plants as in the manufacturing industry. This is also in some way related to long-term cultural changes within the Nuclear Power Industry such as increased production pressures and formalisation of work. In discussing the context of experience development, I have pointed to a wider significance in the use of small-group improvement activities and further possibilities. Naturally, in an international perspective there are many other approaches to quality work and more open, more or less methodical, forms of cooperation than the ones I have presented here in relation to safety-critical work. Hence, undertaking reflection activities on a regular basis is proactive, fostering a reflective praxis or learning organisation. We may categorise it as a problem of leadership or as an outcome of long-term cultural changes of organisations or in general society. If we do not take the time for dialogue and reflection, as it is not counted as achievement, we are likely to inhibit the learning that goes into skill. It is possible that in countries like Japan, with its hierarchical organisations and long working hours, the downgrading of such activities will strike even harder.

What works in one organisation, in one culture, may not necessarily work in another and tsunamis are even more unlikely to pose a threat to power stations in countries like Sweden. Further, it is important that the quality work of a specific organisation is in some sense unique, reflecting the character or "personality" of the organisation and the people working there. The reason why organisational qualities can be a source of competitive advantage is that they are often difficult to mimic. Besides, naive imitation is seldom the way forward. The key numbers and indicators utilised to evaluate operating organisations worldwide will always leave out certain dimensions of quality substantial to their actual functioning. Those approaches to quality work and training discussed here, small-group improvement-type activities and experience development, provide opportunity for professionals to reflect, to develop and evaluate their own practice.[55]

In connection with safety the addition of more rules is hardly the solution. It can even be counter-productive. Long term, reflective practices provide a presence of the unexpected in everyday procedures essential to the maintenance of a dynamic safety culture. In safety-critical activities there must be a sense of individual responsibility and ownership, along with a capacity for critical thinking. Marginalising the human factor by gradually putting in added automation, production pressures or extensive formalisation of work will lead to a decrease in the learning from everyday practice, giving rise to some sort of hollowing out of ability, or de-skilling, such as the enervation of judgement, making practitioners less capable of managing the unexpected. Formalisation of work may not be the obvious source of a long-term deterioration of quality. However, drifting towards an accident will always be difficult to detect.

Notes

1 Cf. Imai, M. (2012): *Gemba Kaizen*, pp. 53–54.
2 Cf. Göranzon, B. and Florin, M. Eds. (1992): pp. 3–4ff; Dekker, S. (2012): *Just Culture – Balancing Safety and Accountability*, p. 73.
3 Rasmussen, J. and Svedung, I. (2007): p. 13.
4 Woods, D., Dekker, S., Cook, R., Johannesen, L. and Sarter, N. (2010): pp. 42–43; 198–199.
5 Ibid., p. 33.
6 See Ratkić, A. (2012): "Images of Reflection: On the Meanings of the Word Reflection in Different Learning Contexts", *AI & Society*, 25th Anniversary Volume.
7 Janik, A., "Literature, Reflection, and the Theory of Knowledge", in Göranzon, B. and Florin, M. Eds. (1991): p. 156.
8 Ibid., p. 157.
9 Cf. Sennett, R. (1999): pp. 132–135.
10 Snow, C. P. (1998): pp. 92–93.
11 Janik, A., "Do The Humanities have a Future?", in Göranzon, B. Ed. (2011): pp. 103–104.
12 Ibid., p. 100.
13 Weick, K. and Sutcliffe, K. (2007): pp. 83–84ff.
14 Janik, A., in Göranzon Ed. (2011): pp. 97–100. Cf. Murray, C. (2008): pp. 119–120.
15 Cf. Göranzon, B., Hammarén M. and Ennals R., Eds. (2006): p. 85f.
16 Cf. Ferguson, N. (2014): p. 63.
17 Backlund, G. and Sjunnesson, J., in Göranzon, B., Hammarén M. and Ennals R., Eds. (2006): p. 135.
18 Hammarén, M., "Skill, Storytelling and Language: on Reflection as a Method", in Göranzon, B., Hammarén M. and Ennals R., Eds. (2006): p. 206.
19 Imai, M. (2012): pp. 8f.
20 Hutchins, D. (2008): *Hoshin Kanri – The Approach to Continuous Improvement*, p. 195.
21 Ibid., pp. 215–216.
22 Weick, K. and Sutcliffe, K. (2007): p. 157.
23 Kay, J. (2011): p. 172.

24 Ibid., pp. 3–4.
25 Woods, D., Dekker, S., Cook, R., Johannesen, L. and Sarter, N. (2010): p. 92.
26 Ibid., p. 91.
27 Imai, M. (2012): pp. 54–57.
28 Hutchins, D. (2008): pp. 144f. Cf. Imai, M. (2012): pp. 95f.
29 Aoki, K., "Transferring Japanese *Kaizen* Activities to Overseas Plants in China", *International Journal of Operations and Production Management*, Vol. 28, No. 6, 2008: p. 533.
30 These arguments were developed at a seminar at the ABF association in Stockholm on the 11 April 2013, on the safety of nuclear power plants in Sweden, arranged by the *Swedish Society for Risk Sciences*.
31 Kurokawa, K. et al. (2012): p. 16. Cf. INPO, "Lessons Learned from the Nuclear Accident at the Fukushima Daiichi Nuclear Power Station" (2012): pp. 9f.
32 Kurokawa, K. et al. (2012): p. 17.
33 INPO, "Lessons Learned from the Nuclear Accident at the Fukushima Daiichi Nuclear Power Station" (2012): p. 28.
34 Ibid., p. 3. As stated in the NIRA report of 1985, in competing with the west many Japanese corporations were keen on making their own applications.
35 Sakuda, H. and Takeuchi, M. (2013): "Safety-critical Human Factors Issues Derived from Analysis of the TEPCO Fukushima Daiichi Accident Investigation Reports", *Nuclear Safety and Simulation*, Vol. 4, Number 2, p. 142.
36 INPO, "Lessons Learned from the Nuclear Accident at the Fukushima Daiichi Nuclear Power Station" (2012): pp. 21–23.
37 Kurokawa, K. et al. (2012): p. 17.
38 Ibid., p. 23.
39 Ibid., p. 28.
40 Ibid., pp. 43–44.
41 "Japan and Abenomics: Riding to the rescue", *Economist*, 8 November 2014.
42 Kurokawa, K. et al. (2012): p. 74.
43 Hutchins, D. (2008): pp. 90–91. This has been implemented at some power plants, like the Sellafield facility in North England, where Quality Circle-type activities have been used for many years.
44 Dekker, S. (2012): pp. 82f.
45 Ibid., p. 159.
46 Sennett, R. (2012): pp. 20–29; 271–277.
47 Yotsumoto, M., Nakanishi, A., Sugihara, D., Takagi, T. and Ushimaru, H., (2014): "Using Same Words, Seeing Different Worlds: A Case of the Fukushima No. 1 Nuclear Accident". Presentation at the Academy of Management, Tokyo.
48 Ibid.
49 Corradini, M. and Klein, D. (2012): p. 29.
50 Kurokawa, K. et al. (2012): p. 33.
51 Seminar at the Research Centre for Human Factors, Institute of Nuclear Safety Systems (INSS), Tsuruga, 19 December 2014.
52 Yotsumoto, M., Nakanishi, A., Sugihara, D., Takagi, T. and Ushimaru, H. (2014): pp. 17–18.
53 Imai, M. (2012): p. 23.
54 Interviews at the Research Centre for Human Factors, Institute of Nuclear Safety Systems (INSS), Tsuruga, 19 December 2014; study visit and interviews at the General Administration Department at Fukushima Daiichi, 6 February 2015.
55 Hutchins, D. (2008): pp. 7–11ff., and Göranzon, B., Hammarén M. and Ennals R., Eds. (2006): pp. 85–173.

7 Conclusions

Leading up to the momentous accident at Fukushima Daiichi, two parallel movements can be delineated in Japanese society: First, the implementation of *Information Society*, a programme by which all manual knowledge was to be formalised and computerised by the year 2000; in part a response to the dominant role of the US at the time. Second, the progression of Quality Circle-type activities in the 1960s and onwards, engaging people in problem solving and proactive thinking. Built on Japanese tradition and its respect for craft, this was also a way of making manufacturers competitive in the global market. When it comes to the Nuclear Power Industry, it seems to have been less fruitful. Yet, arguably, this has less to do with the training itself and more to do with the increased burden of extensive formalisation, and the fact that executives and top management were not committed enough to these aspects of safety and quality. In the view of regulatory authorities, international collaborative bodies and even top management, making a higher degree of schematic knowledge possible is what paves the way for the kind of control and legibility that facilitates benchmarking, risk assessment and auditing. Processes of auditing and formalisation are not neutral, they have various effects on practice and shape organisations. In that sense, long-term cultural changes of work or a migration towards increased formalisation ought to be supplemented and counter-balanced by the bottom-up perspectives of the skilled practitioner, expanding the potential for learning and knowledge development. My experience is this: To some people, typically with certain reviewers and scientists, this is not education. Moreover, it has little to do with safety culture, which to them has everything to do with those aspects of practice that can be tracked down and quantified, in other words the measurement of performance. Still, dichotomies like these can thwart our conceptions of reality, obscuring what lies beyond such taxonomies. Proficient learning is not only linked with changes in observable behaviour or with the assimilation of formal education. In practical activities we also want the professional to be responsible and to learn from one's own mistakes and experiences relative to a given situation. Also, what we can learn from Japanese experience is that standardisation and harmonisation, beyond a certain point, can have a negative impact on the safety culture of high-risk organisations.

The notion of Taylorism reverberates around the separation of thinking and doing, or *knowing* and doing. In recent decades, these tendencies have resurfaced in the west, albeit in new varieties, and have arguably played a part in the slow steady degeneration of western society and culture. Contrary to the Taylorism of old, the objectives of the New Taylorism are those of short-term profits and re-engineering, and usually with the assistance of business consultancy agencies. If we accept that a culture is fashioned by the development of those qualities and faculties that characterise a specific group of people, this will mean that we can also learn from a culture, informally, building on the human capacity for dialogue, cooperation and analogical thinking. Skilful practitioners know more than they can tell, and in a viable work culture there are indirect sources of quality that cannot be replaced by enforced standardisation, or systems of control and auditing. Practitioners possess a great deal of readily available and tacit knowledge which are indispensable assets when facing the next future surprise – experienced operators think forward, into an indeterminate future. Besides, if it is imagined that experience is something that we are able to "transfer" between individuals, and between generations and organisations, this presupposes an insight into the fact that this experience-based knowledge becomes a *new* knowledge among those who "receive" it. The underlying objective of codifying knowledge, from tacit to explicit, highlights an elementary weakness of Knowledge Management also distinguished in Taylorism; approaches to work that have not led to any greater understanding of human skill and labour. A significant implication of Taylorised work places, rather, has been the reduction of knowledge, and thinking, along with a sense of alienation among workers. As in the nuclear industry, the highest quality is replaced by the *right* quality and the obsession of finding a one-to-one match between means and ends. Accordingly, our thinking can be restricted by cultural impacts.

Arguably, many western societies are in a so-called "stationary state" of stagnation, or even degeneration, whereas backwards-oriented forces are gaining more and more publicity and momentum. The stationary state is also fitting to certain companies and organisations: over time they suffer a decline in quality to the point where they deteriorate.[1] In this book I have discussed various sources of degeneration: The separation of thinking and doing; erosion of skill or "de-skilling"; the naive projection of quality; long-term cultural changes of work; as well as shifting priorities of leadership. The Nuclear Power Industry is not an isolated entity. Quite the reverse, there are similarities to other areas of work, not least when it comes to the accumulation of skill and tacit knowledge. If our judgements are by any means clouded, this will ultimately affect the ability to manage the unexpected. There are other causes that may have similar effects. By means of re-engineering and technological change, new varieties of Taylorism are spawned through various industries and, in the public sector, cultural changes of work with ambiguous effects on the development and maintenance of professional skill, as there will be less opportunity for the exertion of personal judgement. At worst we are forced to reduce our judgements and critical thinking. For the probing human mind this represents a de-skilling for the professional. Arguably, lack of autonomy and the separation of thinking and doing are the true enemies of craftsmanship. In other words, the tacit knowledge must "survive", and one should therefore avoid repressing it, which is exactly what is being done in many work place organisations. The skill factor of safety cultures must not be disregarded, as repression of tacit knowledge is generally a source of long-term degeneration. Professional decision making must not proceed against one's better judgement. When daily business activities move in the direction of non-involvement, oriented towards detached knowing-that, payoffs in terms of learning from everyday practice and the progression of skill are likely to significantly wear off, in the course of which the safety culture of high-risk organisations will be increasingly debilitated. Hence, if the process of degeneration is to be counteracted, one must also look at ways of nurturing the skill factor of organisations.

In conclusion, I have tried to demonstrate how, through the impact of automation and enforced standardisation and re-engineering of work, the long-term development and retention of skill is impoverished, as the learning that goes into skill is reduced. As adults' proficiency, in the vein of craftsmanship, is based on slow learning and adaptation, there must be opportunity for reflection and critical thinking, like the personal analysis of mistakes. Practical proficiency gains from digressions and variability, approaching problems and objectives in different ways. One way to confront and counteract contemporary tendencies of significance is through the educational system. Education is an important part of a culture; it is influenced by culture and it is also a means of *influencing* a culture. In safety-critical activities problems will arise where there is little or no counter-balance or alternatives to automation and extensive formalisation. Small-group improvement-type activities or reflection-type activities can provide organisations with a forum for choice and selection on matters of significance, with regard to issues of quality and continuous improvements. This is no substitute for (other) training and education, or work committees, task forces and chain of command, but is a vital supplement. Reflection activities can also be a means of recovering from awkward situations and experiences.

Whereas the accident at Chernobyl in 1986 can be traced back to the deterioration of the Soviet Union, and its varieties of Taylorism, some of the root causes of the Fukushima accident seem to be connected with the accumulation of a culture of compliancy; the safety culture on an organisational level seems to have been lacking a constructive amount of critical thinking. The essence of professional skill is not so much about the management of technology but about the formation of judgements on the basis of analogical thinking; addressing the same or similar problems in different ways. Replacing a knowledge ideal based on analogical thinking with one intimately associated to the concept of a model is likely to have adverse effects on build-up and maintenance of professional skill and judgement. To nurture analogical thinking will empower people with insight into their own experience as well as others, with a key role for dialogue. In safety-critical activities there is arguably a need for more "open" forums of communication and cooperation, to promote mutual exchange and understanding within an organisation or sector of industry – a physical meeting place where discussions can take place but also to create the condition in which experience can meet experience; to capture reality by means of examples rather than models and to make experiences accessible for reflection. In both cases reflection has an impact on knowledge and experience.

Communication in the sense of dialogue is essentially as much about "what is left unsaid as said". Small-group reflection-type activities can thus be a way of bridging the gap between people of the same organisation, as well as the gap between what people know and what they are permitted or able to do. To reflect we sometimes need a *mask*, like a piece of literature, which can illuminate something of the complex character of human experience. The process of reflection does not always come naturally. It often needs to be invigorated by some sort of impulse; a work of preparation. The endeavour is to trigger those involved to reflect, to extend rather than to restrict the critical thinking of the practitioner. Experience development makes way for dialogue and digressions; "zooming out" and opening up for various perspectives to come through, as a way of nurturing mistakes and experiences within a certain praxis.

In high-risk industries there are measures taken to ensure that the learning that derives from training can be efficiently applied and utilised. The executives of nuclear power plants want to be assured that the money spent on training and formal education has its appropriate effects. Preferably they would also want those effects to be measurable, although executives may not themselves have the knowledge they require from others. The reverse also applies: Knowledge informally extracted from real-time situations, mistakes and from a culture must be reflected upon, nurtured and further developed. On a long-term basis, this is what generates the type of skill relevant to manage the undetermined; decision making under opacity. One way of making this happen can be by identifying areas or activities of work in which human capacities can be more fully engaged. In that case recommendations and areas of improvement from external reviewers would be not so much be a push for formalisation as a provider of tasks and puzzles to be evaluated and analysed, as it should be. Some problems are best approached indirectly. Moreover, this is likely to reduce the risk of adjustments to practice merely becoming quick fixes so as to delight supervisory agencies and other outside observers. In contrast to the codification of knowledge promoted in those management strategies commonly regarded as industry best practice, safety-critical work can be rejuvenated by giving practitioners the opportunity to evaluate the quality of their own practice.

Skill reduces risk due to the fact that experienced practitioners have become familiar with a great variety of concrete situations of practice. Compliance with standard procedures is not a venture but can for that matter still be *adventurous*, for the reason that everything cannot be predicted or codified. In other words, discontinuity cannot be ruled out. The world is still unpredictable, maybe even more so owing to the complexity of today's society. Degenerations of quality do not happen overnight; it is the gradual, long-term processes of decline that are likely to pose the greatest threat to the safety culture of high-risk activities. Yet to counteract any such untoward consequences, or a slow steady drift towards accidents, we must be given the tools and opportunity to actually see the problems in the first place before these processes are irreversible and thus cannot be revived.

Note

1 Cf. Ferguson, N. (2014): pp. 137–138.

References

Aoki, Katsuki, "Transferring Japanese Kaizen Activities to Overseas Plants in China", *International Journal of Operations and Production Management*, Vol. 28, No. 6, 2008: 518–539.

Background: The Forsmark Incident 25 July 2006, published by the Analysis Group at The Swedish Nuclear Safety and Training Centre.

Backlund, Göran and Sjunnesson, Jan, "Better Systems Engineering with Dialogue", in Göranzon, Hammarén and Ennals (Eds), 2006.

Berglund, Johan, "Dialogue and Age: Confronting disputed concepts by means of Dialogue", in Ennals, Richard (Ed.), 2009.

Berglund, Johan, "Säkerhet och ekonomisk rationalisering – om informationen till de anställda på svenska kärnkraftverk" (Safety Culture and Economic Rationalisation), Report I: The Swedish Nuclear Safety and Training Centre, 2012.

Berglund, Johan, "Den nya taylorismen – om säkerhetskulturen inom kärnkraftsindustrin" (The New Taylorism: On Safety Culture) PhD thesis, KTH Royal Institute of Technology. Revised edition. Stockholm: Dialoger, 2013.

Berglund, Johan, with Nils Friberg, "Paradox inom kunskapssamhället – om erfarenheter från svensk kärnkraftsindustri" (Paradox of Knowledge Society – experiences from the Swedish nuclear industry) Report II: The Swedish Nuclear Safety and Training Centre, 2012.

"Biological Effects of Fukushima Radiation on Plants, Insects, and Animals", www. phys.org, 14 August 2014, accessed 10 August 2015.

"Comprehensive Study of Microelectronics", National Institute for Research Advancement (NIRA), 1985.

Corradini, Michael and Klein, Dale, "Fukushima Daiichi – ANS Committee Report". The American Nuclear Society Special Committee on Fukushima, March 2012.

Crawford, Matthew B., *Shop Class as Soulcraft – An Inquiry into the Value of Work*. London, Penguin Books, 2010.

Dekker, Sidney, *Just Culture – Balancing Safety and Accountability*. Second Edition. Farnham, Surrey, Ashgate Publishing Limited, 2012.

De Vos, Anne, Lobet-Maris, Claire, Rousseau, Anne, and Wallemacq, Anne, "Knowledge in Question: from Taylorism to Knowledge Management". Paper presented in OKLC 2002, the Third European Conference on Organisational Knowledge, Learning and Capabilities, Athens April 2002, http://www2.warwick. ac.uk/fac/soc/wbs/conf/olkc/archive/oklc3/papers/id479.pdf, accessed 23 September 2013.

Dieckmann, Peter, Gaba, David and Rall, Marcus, "Deepening the Theoretical Foundations of Patient Simulation as Social Practice", *Sim Healthcare*, Vol. 2, No. 3, 2007: 183–193.

Dieckmann, Peter, Molin Friis, Susanne, Lippert, Anne and Østergaard, Doris, "The Art and Science of Debriefing in Simulation – Ideal and Practice", *Medical Teacher*, Vol. 31, No. 7, 2009: 287–294.

Doray, Bernard, *From Taylorism to Fordism. A Rational Madness*. London: Free Association Books, 1988.

Dreyfus, Hubert and Dreyfus, Stuart, *Mind over Machine – The Power of Intuition in the Era of the Computer*. New York: Free Press, 1986.

Engström, Diana, *Att styra säkerhet med siffror* (Managing Safety by Numbers), Master Thesis in Skill and Technology, Linnaeus University, 2015.

Ennals, Richard (Ed.) Special Issue: The Enlightened Workplace, *AI & Society*, Vol. 23, No. 1, January 2009.

Ericson, Mats, and Mårtensson, Lena, "The Human Factor?", in Grimvall, G., Homlgren, Å, Jacobsson, P. and Thedéen, T. (Eds), 2010.

Ferguson, Niall, *The Great Degeneration – How Institutions Decay and Economies Die*. London: Penguin Books, 2014.

Göranzon, Bo (Ed.) *Skill, Technology and Enlightenment: on Practical Philosophy*. London: Springer Verlag, 1995.

Göranzon, Bo, "Tacit Knowledge and Risks", in Göranzon, B., Hammarén, M. and Ennals, R. (Eds), 2006.

Göranzon, Bo, (1990) *The Practical Intellect – Computers and Skills*. Rev. Ed., Stockholm: Santérus Academic Press, 2009.

Göranzon, Bo, and Florin, Magnus (Eds) *Dialogue and Technology – Art and Knowledge*. London: Springer Verlag, 1991.

Göranzon, Bo, and Florin, Magnus (Eds) *Skill and Education: Reflection and Experience*. London: Springer Verlag, 1992.

Göranzon, Bo, Hammarén, Maria and Ennals, Richard (Eds), *Dialogue, Skill and Tacit Knowledge*. London: John Wiley, 2006.

Grimvall, Göran, Holmgren, Åke J., Jacobsson, Per and Thedéen, Torbjörn (Eds) *Risks in Technological Systems*. London: Springer Verlag, 2010.

Groopman, Jerome, "Diagnosis: What Doctors Are Missing", *New York Review of Books*, November, 2009: 5–18.

Gustafsson, Lars and Mouwitz, Lars, "Validation of Adults' Proficiency – Fairness in Focus", National Centre for Mathematics Education, NCM, University of Gothenburg, 2008.

Hammarén, Maria, "Skill, Storytelling and Language: On Reflection as a Method", in Göranzon, B., Hammarén, M. and Ennals, R. (Eds), 2006.

Head, Simon, "Inside the Leviathan", *New York Review of Books*, December 16 2004.

Head, Simon, "They're Micromanaging you're Every Move", *New York Review of Books*, August 16 2007.

Hughes, William, *Critical Thinking: An Introduction to the Basic Skills*. Second Edition. Peterborough, Ont: Broadview Press, 1996.

Hutchins, David, *Hoshin Kanri – The Strategic Approach to Continuous Improvement*. Aldershot, Hampshire: Gower Publishing Limited, 2008.

Imai, Masaaki, *Gemba Kaizen – A Commonsense Approach to a Continuous Improvement Strategy*. Second Edition. New York: McGraw-Hill, 2012.

Janik, Allan, "Literature, Reflection and the Theory of Knowledge", in Göranzon, B. and Florin, M. (Eds), 1991(a).

Janik, Allan, *Cordelias tystnad – om reflektionens kunskapsteori* (The Silence of Cordelia – on the Epistemology of Reflection). Stockholm: Carlssons, 1991(b).

Janik, Allan, "Do the Humanities have a Future?", in Göranzon, B. (Ed.) Turingmänniskan (Turing's Man). *Dialoger* Nos. 87–88, 2011.

"Japan and Abenomics: Riding to the rescue", *Economist*, 8 November 2014.

Johannessen, Kjell S., "Rule Following, Intransitive Understanding and Tacit Knowledge: An Investigation of the Wittgensteinian Concept of Practice as Regards Tacit Knowing", in Göranzon, B., Hammarén, M. and Ennals, R. (Eds), 2006.

Josefson, Ingela, "A Confrontation between Different Traditions of Knowledge – An example from Working Life", in Göranzon, B. (Ed.) 1995.

Josephs, Herbert, *Rameau's Nephew: A Dialogue for the Enlightenment*, in Göranzon, B. and Florin, M. (Eds), 1991.

Kay, John, *Obliquity – Why Our Goals Are Best Achieved Indirectly.* London: Profile Books, 2011.

Kurokawa, Kiyoshi et al., The National Diet of Japan Fukushima Nuclear Accident Independent Investigation Commission, NAIIC – The Official Report of the Fukushima Nuclear Accident Independent Investigation Commission (Executive Summary), 2012.

Larsson, Lars G., and von Bonsdorff, Magnus, "Ledarskap för säkerhet – Insikt, engagemang, förändring: En oberoende utredning av säkerhetsfrågornas hantering inom Vattenfalls kärnkraft i Sverige" (Leadership for Safety: An independent report on the management of Safety within Vattenfall Nuclear Power), 2007.

Lave, Jean and Wenger, Etienne, *Situated Learning – Legitimate Peripheral Participation.* Cambridge: Cambridge University Press, 1991.

"Lessons Learned from the Nuclear Accident at the Fukushima Daiichi Nuclear Power Station". INPO 11–005, Rev. O, Addendum, Special Report, Institute of Nuclear Power Operations, August 2012.

Murray, Charles, *Real Education – Four Simple Truths for Bringing America's Schools Back to Reality.* New York: Three Rivers Press, 2008.

Nordenstam, Tore, "Technocratic and Humanistic Conceptions of Development". Research Report No. 51, The Swedish Centre for Working Life, 1985.

Nordenstam, Tore, *The Power of Example.* Stockholm: Santérus Academic Press, 2009.

O'Doherty, John P., "Decisions, Risk and the Brain", in Skinns, L., Scott, M. and Cox, T. (Eds), 2011.

Perby, Maja-Lisa, *Konsten att bemästra en process – om att förvalta yrkeskunnande* (The Art of Mastering a Process – On the Management of Skill). Gidlunds: Hedemora, 1995.

Perin, Constance, *Shouldering Risks: The Culture of Control in the Nuclear Power Industry.* Princeton, NJ and Oxford: Princeton University Press, 2007.

Perrow, Charles, *Normal Accidents: Living with High-Risk Technologies. With a New Afterword and a Postscript on the Y2K problem.* Princeton, NJ: Princeton University Press, 1999.

Perrow, Charles, *The Next Catastrophe: Reducing our Vulnerabilities to Natural, Industrial and Terrorist Disasters.* Princeton, NJ: Princeton University Press, 2011.

"The Plan for Information Society – a National Goal toward Year 2000". Japan Computer Usage Development Institute, Computerization Committee, Final Report, May 1972.

Polanyi, Michael, (1966) *The Tacit Dimension*. Gloucester, MA: Peter Smith, 1983.

Power, Michael, *The Audit Society – Rituals of Verification*. Oxford: Oxford University Press, 1997.

Power, Michael, *Organized Uncertainty – Designing a World of Risk Management*. Oxford: Oxford University Press, 2007.

Rasmussen, Jens, and Svedung, Inge, *Proactive Risk Management in a Dynamic Society*. Second edition. Karlstad: National Centre for Learning from Incidents and Accidents, 2007.

Ratkić, Adrian, "Images of Reflection: On the Meanings of the Word Reflection in Different Learning Contexts", *AI & Society*, 25th Anniversary Volume, www.springerlink.com, 2012.

Sakuda, Hiroshi, and Takeuchi, Michiru, "Safety-critical Human Factors' Issues Derived from Analysis of the TEPCO Fukushima Daiichi Accident Investigation Reports", *Nuclear Safety and Simulation*, Vol. 4, No. 2, June 2013: 135–146.

Sandblad, Anders, "Maskiner och människor" (Man and Machines), Master Thesis in Skill and Technology, Linnaeus University, May 2015.

Scott, James C., *Seeing Like a State: How Certain Schemes to Improve the Human Condition Have Failed*. New Haven, CT and London: Yale University Press, 1998.

Sennett, Richard, *The Corrosion of Character – The Personal Consequences of Work in the New Capitalism*. New York, NY and London: W.W. Norton, 1999.

Sennett, Richard, *Respect in a World of Inequality*. New York, NY & London: W. W. Norton, 2004.

Sennett, Richard, *The Culture of the New Capitalism*. New Haven, CT and London: Yale University Press, 2006.

Sennett, Richard, *The Craftsman*. New Haven, CT and London: Yale University Press, 2008.

Sennett, Richard, *Together – The Rituals, Pleasures and Politics of Cooperation*. New Haven, CT and London: Yale University Press, 2012.

Skinns, Layla, Scott, Michael, and Cox, Tony (Eds) *Risk. The Darwin College Lectures*; 24. Cambridge: Cambridge University Press, 2011.

Snow, C. P., *The Two Cultures*, with Introduction by Stefan Collini. Cambridge: Cambridge University Press, 1998.

"Social Change in Japan: When the Myths are Blown Away", *Economist*, 19 August 2010.

Spiegelhalter, David, "Quantifying uncertainty", in Skinns, L., Scott, M. and Cox, T. (Eds), 2011.

Spelplats 1.2009. Kärnkraftsäkerhet och utbildning. Margaretha Engström (Ed.), Dialoger.

Spelplats 3.2009. Kvalificerad erfarenhetsutveckling – inom säkerhetskultur och utbildning. Johan Berglund (Ed.), Dialoger.

Spelplats 3.2010. Säkerhetskultur. Susanne Hamlin (Ed.), Dialoger.

Spelplats 1.2011. Om lärande i säkerhetskritiska miljöer. Nina Brandting (Ed.), Dialoger.

Spelplats 1.2012. Den mänskliga faktorn. Johan Berglund and Anette Leijonberg (Eds.), Dialoger.

Spelplats 2.2014. Vad är kvalitet? Liselotte Herlitz (Ed.), Dialoger.

Spelplats 3.2014. Säkerhetskultur över organisationsgränser. Mona Eriksson (Ed.), Dialoger.

Spelplats 4.2014. Kärnkraftsbranschens framtid. Örjan Eklöf (Ed.), Dialoger.

Stensson, Patrik, "The Quest for Edge Awareness: Lessons Not Yet Learned". PhD Thesis. Uppsala University, 2014.

Stewart, James R., "The Work Ethic, Luddities and Taylorism in Japanese Management Literature". *Industrial Management*, 1 November 1992, http://www.thefreelibrary.com, accessed 15 November 2013.

Taleb, Nassim Nicholas, *The Black Swan: The Impact of the Highly Improbable*. Second Edition. New York: Random House, 2010.

Taleb, Nassim Nicholas, *Antifragile: Things that Gain from Disorder*. London: Penguin Books, 2013.

Taylor, Frederick W. (1911) *The Principles of Scientific Management*. Filiquarian Publishing, LLC, 2007.

Weick, Karl E. and Sutcliffe, Kathleen M., *Managing the Unexpected: Resilient Performance in an Age of Uncertainty*. Second edition. London: John Wiley, 2007.

Woods, David D., Dekker, Sidney, Cook, Richard, Johannesen, Leila, and Sarter, Nadine, *Behind Human Error*. Second Edition. Aldershot, Hampshire: Ashgate Publishing Co., 2010.

Yotsumoto, Masato, Nakanishi, Aki, Sugihara, Daisuke, Takagi, Toshio and Ushimaru, Hajime, "Using Same Words, Seeing Different Worlds: A Case of the Fukushima No. 1 Nuclear Accident". Presentation at Academy of Management, Tokyo, January 2014.

Index

Printed and bound by CPI Group (UK) Ltd, Croydon, CR0 4YY

24/10/2024

01778283-0018